I0071665

What in the World are Electrolytes?

Written by:
Austin Mardon, Alexandra Hauser, Alyssa Wu,
Anna Yang, Amir Ala'a, Angel Xing, Amal Rizvi,
Alexa Gee, Alexia Di Martino, Amna Zia,
Amna Abu Askar, & Ami Patel

Edited by:
Jenny Kang

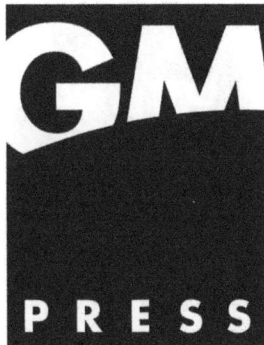

GM

PRESS

Copyright © 2021 by Austin Mardon
All rights reserved. This book or any portion thereof may not be reproduced or used in any manner whatsoever without the express written permission of the publisher except for the use of brief quotations in a book review or scholarly journal.
First Printing: 2021

Typeset and Cover Design by Kim Huynh

ISBN 978-1-77369-236-4
Golden Meteorite Press
103 11919 82 St NW
Edmonton, AB T5B 2W3
www.goldenmeteoritepress.com

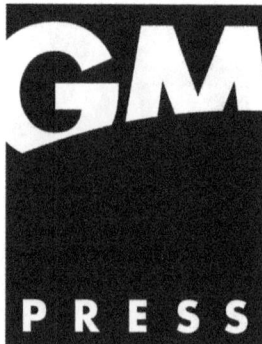

TABLE OF CONTENTS

CHAPTER 1

History of Electrolytes

Alexandra Hauser

Electrolytes are substances that can be studied through many scientific lenses. From physics, to chemistry, to biology, electrolytes are found in journals and textbooks of many branches of science. As a generalized definition of the term, electrolytes are substances with the ability to conduct electricity due to their dissociation into positively and negatively charged particles, which migrate to and are normally discharged at negative and positive terminals of an electric current (The Editors of Encyclopaedia Britannica, 2017). To find value in this working definition of electrolytes, the history of our understanding of electrolytes must be looked at.

Generally speaking, electrolytes have been recognized for over a century, and are very closely tied to the development of the field of electrochemistry. One of the first experiments in regard to electrochemistry was completed by Luigi Galvani in 1786, while studying the effects of atmospheric electrical discharge (lightning). He observed that the legs of a recently dissected frog would twitch when it created a circuit between two different kinds of metal (MIT Libraries, 2021). To understand what was going on, Galvani attached the frog legs to a brass hook and iron railing during a thunderstorm. He found that the legs twitched spastically both when there was lightning and when the sky was calm and called this observation "animal electricity" (Surat, 2019). These observations and experiments prompted the concept of 'Galvanism,' which is the movement of muscles in response to electricity.

Objecting to "animal electricity," Alessandro Volta proposed that the results came from the use of two dissimilar metals connected by a moist conductor (the frog's leg) (The Electrochemical Society, 2021). To prove this, he developed a circuit consisting of two different metals separated by a piece of cloth soaked in saltwater. By stacking these circuits (pairs of copper and zinc disks with an electrolyte (cloth soaked in saltwater) in between) Volta was able to control the amount of electricity produced, and developed the first battery, called a 'voltaic pile' (MIT Libraries, 2021). After his invention, Volta proceeded to publish one of the earliest electromotive series, which ranked metals and other substances based on the strength of their electrical effects. This series placed metals that would create the strongest effects furthest apart and was very similar to the tables other scientists were developing that showed what substances would displace others in compounds (The Electrochemical Society, 2021).

Around this time, the concept of electrolysis, or the breaking of an element into its constituents through the addition of an electric current, was discovered. Nicholson and Carlisle read about Volta's voltaic pile and decided to construct one of their own using crowns (a British coin made of pure silver), zinc, and pasteboard that was soaked in saltwater (Mander, 2017). They then placed a drop of river water on the uppermost disk and found that hydrogen gas had been released. Hypothesizing that the hydrogen gas was due to the decomposition of the water via the electric current, they created a new apparatus with the voltaic pile where the brass wires flowed through a tube filled with water. Upon observation, there were small bubbles rising from the tube, and the bottom wire attached to the silver blackened due to oxidation (Mander, 2017). They believed that the larger stream of bubbles rising was hydrogen, and the smaller stream was oxygen, meaning that they were correct in that the electric current had broken down the water into its constituent elements (Mander, 2017).

Jöns Berzelius continued the research into electrolysis and in 1803, alongside Wilheim Hisinger, found that not only could electric current decompose water, but the constituents each accumulate at a different pole. They also found that it could separate solutions of salts so that acids moved to one pole and alkalies moved to the other (Jorpes, 1970). This allowed him to assume that the electric current was making the molecules split into positive and negative parts, and that the neutrality of molecules was due to the mutual neutralization of the opposite charges (Science History Institute, 2017).

Around the same time, after hearing about the voltaic pile, Humphrey Davy determined that the production of electricity depended on a chemical reaction taking place and continued with electrochemical experiments that allowed him to find that the tendency for substances to react with each other is electric in nature (Science History Institute, 2017). Then, based on the concept of electrolysis, in 1807, Humphrey Davy electrolyzed various compounds, (breaking them down into their constituents) and discovered many new elements (ie. sodium, potassium, calcium, etc.), and came to the conclusion that an electrical force keeps different elements together in compounds (Surat, 2019).

In 1833, Michael Faraday began to study electrolysis. On a circular glass plate, there were two square tin foils and between the foils there was a circular filter paper soaked with a chemical solution. The left and right foils were connected to the positive and negative sides of a battery. Platinum wires with special shapes were connected to the filter paper, connecting the circuit. Depending on the chemical solution used there were different results (Faraday & Royal Society (Great Britain), 1833). For example, filter paper soaked in a solution with potassium iodide and starch had a blue spot appear under the positive wire from the production of iodine. Using the different chemicals allowed for Faraday to develop the laws of electrolysis that state that the amount of products generated is proportional to the amount of charge passed through the circuit and that the amounts of chemical changes produced by the same quantity of electricity in different substances is proportional to their equivalent weights (The Editors of Encyclopaedia Britannica, 2021).

In 1834, Faraday published these results with the inclusion of new terminology. He said the word 'pole' insinuated attraction or repulsion from the poles themselves, rather than the poles being extremities of charge caused by attraction and repulsion. So, Faraday introduced the term electrodes (Faraday & Royal Society (Great Britain), 1833). Instead of using 'positive electrode' and 'negative electrode,' Faraday used the term anode to refer to the negative extremity of the decomposing compound, and cathode to refer to the positive extremity. Of all the new terminology, the decision to call the decomposing compound of which elements are set free electrolytes is the most important. Rather than calling the ions (elements that are set free) electro-positive and electro-negative, he combined the electrode they moved towards with their original term, making ions that move towards the cathode, cations, and ions that move towards the anode, anions (Faraday & Royal Society (Great Britain), 1833). To put this in simpler terms, lead chloride is an electrolyte, which when electrolyzed, decomposes to two ions, the chlorine being the anion and lead being the cation.

Over the next 30 years, many different concepts came together. One being the ideal gas law, which is the combination of Boyle's law, Charles' law, and Avogadro's law, and describes various information, including pressure, volume, temperature, and number of moles, for a perfect gas. One of the more notable conclusions in regard to electrolytes is that the pressure of a perfect gas is directly proportional to concentration, given that volume and pressure are constant (LeTran & Chemistry LibreTexts, 2020). In 1877 Wilhelm Pfeffer researched osmosis, the movement of water across the cell membrane, and developed a method to measure osmotic pressure (The Editors of Encyclopaedia Britannica, 2021). Pfeffer showed that osmotic pressure depended on molecules that were too large to pass through the membrane pushing against the cell wall/membrane. The protoplasm (cytoplasm) of the cell pushes back, and the cell wall reacts to the protoplasmic pressure with an equal and internally directed pressure, leaving a turgor pressure (more pressure inwards than outwards) (Parker, 2017). As well, Pfeffer worked with different dilutions of water and the large solute, as well as developed a temperature coefficient of osmotic pressure (Jones, 1899). Another concept was by Jacobus van't Hoff who said that osmotic pressure is directly proportional to the concentration of the solution and can be modeled by the ideal gas law (Ramberg, 2021).

In 1887, Svante Arrhenius developed his theory that suggested that when an acid, base, or salt is dissolved in water it will undergo spontaneous dissociation, forming a positive and negative ion (Jones, 1899). Using the three concepts mentioned before, Arrhenius looked at the osmotic pressure of certain solutions, those solutions being the combination of strong acids, bases, and salts dissolved in water. He found that the osmotic pressure exerted by the solutions was greater than the pressure calculated. It could not be due to an error in combination of the concepts, as he carried out the same experiment with solutions that could not conduct electricity (ie. solutions of cane sugar and water) (Jones, 1899), therefore there must be another explanation for the inconsistencies. Following the past anomalies that were discovered with the gas law, Arrhenius determined that the increase in osmotic pressure was due to the acids, bases, and salts breaking down into their constituents, and thus they dissociate into positive and negative ions in water (Jones, 1899). Arrhenius actually developed this theory in 1884 when he wrote a paper "On the Electrical Conductivity of Electrolytes," but due to severe objections raised by peers, mainly concerning the reasonability of the theory, he instead wrote about active and inactive parts of a molecule (De Berg, 2003).

At the time, Arrhenius' dissociation theory was not fully accepted by the scientific community, but in 1884, Wilheim Ostwald received the doctoral thesis of Arrhenius which stated his theory that was mentioned before. Ostwald recognized that if all acids contained the same ions, then the differences in chemical activity was a result of the concentration of the ions (Schummer, 2021). He then took on establishing a dissociation law that was published in 1888 that uses the degree of dissociation and concentration of the acid to determine its equilibrium constant. The law was tested by measuring the electrical conductivities of more than 200 acids, and helped to support Arrhenius' dissociation theory (Schummer, 2021).

Major connections between the dissociation theory and thermodynamics were made by Walther Nernst, who was inspired by Arrhenius. After developing a way to calculate a diffusion coefficient for electrolytes in infinitely dilute solutions, Nernst established relationships between ionic mobility, diffusion coefficients, and the electromotive force in concentration cells (Kormos-Buchwald, 2020). He perfected the theory of galvanic cells (a type of electrochemical cell that utilizes electrolytes to undergo spontaneous redox reactions, producing an electrical current) by assuming that there was an electrolytic pressure from the dissolution of the electrolyte. This pressure forces the electrodes to release ions into the solution to oppose the pressure from the electrolyte ions that have been dissociated into the solution (Nobel Media AB, 2021). Nernst also discovered the liquid junction potential which occurs when solutions of electrolytes of different concentrations encounter each other. The electrolyte dissociates into ions which migrate to either the low or high concentrated side, and begins an electromigration of ions (Huebener, 2013). In his habilitation thesis in 1889, Nernst presented an equation that represented the fundamental connection between thermodynamics and dissociation theory by describing the electrode potential in electrolytes, or the equilibrium potential of an electrode. It uses the electrode's standard potential and the concentrations of the reactant and product to determine the electrode's overall potential (Huebener, 2013).

Towards the end of the nineteenth century, the field of electrolytes was gaining more interest. From Arrhenius' dissociation theory to the work completed by Ostwald and Nernst, many chemists were doing research in the field of physical chemistry, including Paul Walden, who was once a student of Ostwald. Walden studied the electrical conductivity of salts (electrolytes) of polyacidic bases (Boeck, 2019). From this, he determined absolute values for the charges of cations and anions that dissociate from a strong electrolyte due to observing the increase in equivalent conductivities of solutions having 1/32 and 1/1024 equivalent concentrations and developed a rule from it called the Ostwald-Walden-Bredig rule. This rule has little presence in the 21st century but played a large role at the time as it allowed for the determination of the stoichiometry of salts (Boeck, 2019).

For some time, Walden moved to the field of organic chemistry, but eventually returned to physical chemistry. Upon his return, Walden proposed a question that had great influence: whether only water could dissociate electrolytes (Boeck, 2019). Ostwald's research focused on water, with only five of his calculations including non-aqueous solvents. In 1899, Walden began experimenting with non-aqueous solutions and their ability to dissociate electrolytes into ions. During this research it is believed that Walden developed the idea of solvation, which is the interaction of a solvent with a dissolved solute, in most cases with water, it is referring to hydration (Speight, 2020). From his research, Walden concluded that electrolytes could have electrical conductivity in non-aqueous media too, provided that they are soluble and able to dissociate. He found that the closer the media is to water (ie. it is polar, and has a high dielectric constant), the equivalent conductivities increase at lower concentration until it reaches a maximum. By multiplying this maximum by the viscosity of the solvent, Walden found a constant which he termed Walden's rule, that is approximately constant for the same ions in different solvents (Boeck, 2019; Oxford Reference, 2021).

Paul Walden also tried to find an explanation for the degree that a solvent dissociates an electrolyte. Using Nernst's assumption that the dielectric constant is definite, Walden determined the dissociation ability of a solvent using the degree of dissociation. Through various experiments, he found a relationship between the dielectric constant and volume of different solvents where they possessed the same degree of dissociation of an electrolyte (Boeck, 2019).

Another notable aspect of Walden's research during this time is his experiment with two different salts and various non-aqueous solvents. Comparing the degrees of dissociation for the salts across each solvent, they were very similar for water and alcohol but began to vary more and more across acetone, methylene chloride, and chloroform (Hall, 1929). The cause of this was determined to be the behaviour of strong acids which dissociate nearly completely in water and alcohol and become weaker and weaker electrolytes as less basic solvents are used.

Petrus Debye and Erich Hückel were the next to add to what is known about electrolytes. In the 1920s they developed a theory that allowed them to create universal laws which govern the salt activity coefficients (Kunz, 2014). They found that the properties of electrolyte solutions drastically deviated from laws present at the time, which were based on Van der Waals interactions, and that electrolyte solutions have electrostatic forces that are governed by Coulomb's law (Malley, Artika, & Chemistry LibreTexts, 2020). Using the charges on the anion and cations as developed by Nernst, as well as the ionic strength, relative dielectric constant, and temperature of the solution, they created an equation that allowed for the calculation of the activity coefficient, which accounts for deviations from ideal behaviour in thermodynamics and describes an intermolecular interaction within a solution (Bradley, Kammermeier, & Chemistry LibreTexts, 2020). The ionic strength is limited by the Debye-Hückel limiting law and is calculated using the molality and charge of the ith ion of the electrolyte (Malley, Artika, & Chemistry LibreTexts, 2020). This theory assumes that the electrolytes in solution are fully dissociated and while it is a good description of dilute solutions of strong electrolytes, it is less useful for more concentrated solutions (Oxford Reference, 2021).

Taking the concept that electrolytic dissociations can occur without a change in colour, in 1909, Niels Bjerrum proposed an alteration to Arrhenius' dissociation theory. After completing experiments where multiple electrolytes were dissociated into acids, Bjerrum determined that with strong electrolytes, the change in colour is independent from concentration of acid, and with weak electrolytes, the change in colour changes with the concentration of acid (Bjerrum, 1909). Bjerrum proposes that the change with colour in weak electrolytes is due to the ions entering into a combination with each other, and the dissociation not being complete. The combination being oppositely charged ions that associate into dimers, forming Bjerrum pairs (Adar, Markovich & Andelman, 2017). He believed that the salt that is not dissociated is corresponding in structure to complex salts and that its presence causes changes in the colour and decreases the conductivity of the solution (Bjerrum, 1909). Bjerrum believed that the decrease in conductivity was not due to there being less ions but because the ions moved more slowly since there is a smaller amount of electrolytic friction: positive and negative ions collide with each other less frequently and having more ions will increase the electric field around the ions, creating and thickening a water-mantle. In his new theory, Bjerrum states that the degree of dissociation for strong electrolytes will be "approximately determined by the valency of the ions (Ostwald-Walden's rule), by the dielectric constant of the solvent (Walden), and by the concentration of the salt" (Bjerrum, 1909). The degree of dissociation would rely on electric constants because the complete dissociation of the ions would decrease the electrolytic friction, as opposed to what was mentioned earlier, (decrease in conductivity because they had a decrease in friction) due to the electrostatic forces from the ions. Bjerrum's hypothesis would explain why there are various anomalies with the degree of dissociation of strong electrolytes, including how the law of mass action does not apply.

In 1932, Lars Onsager and Raymond Fuoss extended the Debye-Hückel theory by determining laws of diffusion and conduction for multi-component systems. They looked at the effects of Coulomb forces on transport processes, such as diffusion, electrolytic conduction, and viscous flow (Onsager & Fuoss, 1932). Onsager and Fuoss found that in the case of simple electrolytes, the electrostatic effects are proportional to the square root of the concentration, and derived general limiting laws for conduction in mixtures.

Statistical mechanics was initially developed in the 1870s with the work of Ludwig Boltzmann on the kinetic theory of gases and was termed in 1876 by James Maxwell (Ebeling & Sokolov, 2005). It was also influenced by Josiah Gibbs, and statistical physics was completed by Albert Einstein in 1905 (Bergersen, 1998). Essentially, statistical mechanics is a branch of physics that combines principles and concepts from statistics with those of quantum and classical mechanics, usually with regards to the field of thermodynamics (The Editors of Encyclopaedia Britannica, 2011). It aims to predict measurable properties and qualities of systems based on its constituents. Statistical mechanics is particularly useful when it comes to fluids by using probability theory to "determine the distribution of molecular motions and states in a many-molecule system and provides a method to average the states to obtain the macroscopic (bulk) properties" (Smith, Inomata & Peters, 2013). In the 1950s, statistical mechanics was introduced in the research on electrolytes as it allowed for chemists to propose alternatives based on integrals instead of differential equations (Kunz,

2014). The alternatives based on integrals allowed for the validity of electrolyte activity coefficients to be extended to more concentrated solutions, and eventually led to Kenneth Pitzer's extension of the theory to solutions of high concentrations (Kunz, 2014).

As science modernized, computers allowed for the simulation of electrolyte interactions, allowing for an incredible amount of gain in understanding of properties of electrolytes. Developments in x-ray and neutron scattering experiments have lent to a greater understanding of ion-ion and ion-solvent interactions (Kunz, 2014). Advanced spectroscopic techniques have also helped with these developments, and better models have been able to be made with the combination of the experimental research and computer simulation which have led to even more development with ions near electrolyte interface layers (Kunz, 2014).

Electrolytes have a long and winded history. Since the development of Arrhenius' theory of dissociation in 1887, which continues to be the most modern and widely accepted version of electrolyte theory, many chemists and physicists have been working to help with our understanding of these interesting molecules. With the development of computer simulation and other modern experimental technologies, the interactions of electrolytes and their properties are better understood than ever before. With the continuation of research on electrolytes, a topic where chemistry, physics, and biology overlap, we will have an even greater comprehension of these molecules that play a large role in our everyday lives.

CHAPTER 2

Impact on the Human Body

Alyssa Wu

Fluid and electrolyte balance is an area in health that is commonly misunderstood by the general public. It is important to maintain good cardiovascular health by balancing sodium and potassium intake levels to reduce morbidity and mortality. Research shows that consuming a healthy diet containing high amounts of fruits and vegetables can be protective in maintaining electrolyte balances (Bennett et al., 2020). In the metabolic care of critically-ill patients, it is important for healthcare professionals to help these patients maintain good electrolyte balances to support the function of their gastrointestinal tract and overall nutritional health (Lobo, 2007).

Insufficient pH blood concentrations can have adverse effects on human health. During the process of metabolism and cellular respiration, electrolytes facilitate the absorption of fluid and nutrients from the gastrointestinal tract (Chowdhury & Lobo, 2015). As electrolytes are charged molecules, they induce a net charge that influences plasma pH (Kaplan & Kellum, 2010). This characteristic of electrolytes can be implemented in many therapeutic interventions, which will be discussed later in this chapter and throughout this book.

Cardiovascular disease risk
Obesity has been a major global healthcare concern, which has affected over 50% of Americans. Obesity is also a major risk factor for other cardiovascular diseases (CVD), including coronary artery disease, myocardial infarction (heart attack), hypertension, stroke, and diabetes mellitus type 1 and type 2 (Powell et al., 2000). It is difficult to test for the onset of CVD in obese patients as there are weight limitations with current diagnostic tests, including fluoroscopic, computed tomography and lithotripsy units (Powell et al., 2000).

In patients with cardiac arrhythmias, different ion concentrations (including potassium, calcium, and sodium) have electrophysiological effects on the myocardial fibers in the heart (Surawicz, 1966). Cardiac arrhythmias are most heavily related to the influx and metabolism of potassium ions due to the electrophysiological properties of the transmembrane resting potential (TRP) within the heart muscles (Fisch, 1973). In a study comparing the effects of dietary potassium intake in hypertensive and non-hypertensive patients, potassium has shown to be an important regulator in lowering blood pressure. In patients with hypertension, supplementing the diet with potassium can be used as an alternative to consuming antihypertensive medication (Adrogué & Madias, 2007). By using potassium to reduce blood pressure, this reduces the risk of the onset of other cardiovascular events, including stroke, coronary heart disease, myocardial infarction, and heart attack (Houston, 2011). On the contrary, high levels of sodium have been shown to increase blood pressure (Adrogué & Madias, 2007; Watson et al., 1980). It is important to note that early prevention of hypertension is critical to overcoming the lifestyle implications that are associated with the onset of this cardiovascular disease. Potassium and sodium levels in the diet are linked to the primary prevention and treatment of hypertension (Adrogué & Madias, 2007). High salt (sodium chloride, NaCl) and potassium diets reduce systolic and diastolic blood pressure, which puts individuals at a higher risk of becoming hypertensive (Iqbal et al., 2019). Studies have shown that calcium has beneficial effects that protect an individual from developing gestational hypertension during pregnancy (Imdad et al., 2011; Khanam et al., 2018). Interestingly, individuals who consume diets marketed for "weight loss" have a lower caloric density and high potassium levels. This process leads to an increase in metabolic rate which improves weight loss (Knochel, 1989). A study done by Pahl et al., 1988 looked at the effect of weight loss when a group of 46 moderate to severe obese individuals were on a supplemented fasting health intervention. The researchers used chemical biomarkers, such as serum concentrations of electrolytes, glucose, cholesterol, triglycerides and creatinine. They tracked weight loss throughout the duration of the interventional study, but found that weight loss was most significant during weeks 4-6. They associated this finding with the significant decrease in serum cholesterol and triglyceride levels.

Electrolyte transport therapy has been proposed as a plausible solution to helping patients manage the symptoms associated with cardiovascular disease. In clinical settings, electrolyte transporters are used in cardiology to adjust the metabolic disturbances of the myocardium (heart muscle), coronary circulation disorders, and for patients with hypoxia, which is a condition linked to insufficient oxygen supply at the tissue level (Nieper & Blumberger, 1966).

Healthcare and aging

Aging is a universal process that affects all humans and animals around the world. There are many health conditions associated with aging, such as neurological diseases and physiological deterioration of the bodily systems. In present time, there is an immense amount of research set to extend the upper limit of the human lifespan (Geokas et al., 1990). In a surgical healthcare setting, it is important for physicians to follow a standardized procedure of care for patients undergoing surgery. There are four main stages of perioperative care: ward admission, anesthesia, surgery, and recovery. Proper care should be given to patients managing perioperative pain, as this process activates the autonomic nervous system and can cause some adverse effects on various organ systems (Bradshaw et al., 1998).

Electrolytes affect water balance by altering homeostatic mechanisms in the human body. In our global senior population, the normal physiological changes associated with aging influences the imbalances between fluids and electrolytes (Luckey & Parsa, 2003). As humans age, our kidney functioning tends to diminish over time. It has been hypothesized that approximately 50% of seniors over aged 75 years old are believed to have kidney disease (Aging and Kidney Disease, 2014). In the aging process, it is normal for the body to progressively decline in the function of tissues and organ systems. The kidney is a major organ found in the renal system, whose function is to filter and clear out excess waste products from the body. The kidney filter is made up of three layers, which include endothelial cells, the glomerular basement membrane and epithelial cells, known as podocytes (Rinschen et al., 2015). This filtration barrier clears out excess toxins from the body by restricting the passage of larger molecules based on their relative size, shape and charge (Pavenstädt et al. 2003). As kidney function slowly deteriorates, the glomerular filtration rate (GFR) steadily decreases. GFR measures how the kidney filters blood throughout the body. GFR is also used as a biomarker to detect the potential onset of chronic kidney disease (Stevens et al., 2006). Other biomarkers used to assess overall kidney function include creatinine, glucose, total protein, and β_2-microglobulin (Zamora et al., 1998). Other physiological changes associated with electrolyte imbalances include an increase in atrial natriuretic peptide (ANP) and decreases in urinary concentrating ability, aldosterone, the thirst mechanism, and free-water clearance. The discussion on declining kidney function on aging will be further discussed in the next subsection as well.

The effect of electrolytes on the renal system
Acid-base imbalances can lead to renal failure (Marx et al., 2019). As mentioned in a previous section, one of the main roles of electrolytes is in the absorption of nutrients and fluids from the intestinal system. Electrolytes influence the formation of urinary stones, which can be found in the kidney, bladder and intestine. The common risk factor among patients with urinary stones is obesity, which presents a challenge to physicians and surgeons. Laboratory serum chemistry tests have shown that some biomarkers in obese individuals include higher glucose concentrations, uric acid and parathyroid hormone (PTH). In addition, obese patients also have lower levels of albumin and electrolytes including calcium and magnesium, when compared to the non-obese patients (Powell et al., 2000).

The kidney is a critical organ that is part of the renal system. It is primarily responsible for regulating the electrolyte and fluid status of the human body (Nanovic, 2005). Renal complications are most often associated with three main areas of the kidney, including the glomeruli, tubules and vascular compartments (Carriazo et al., 2020). Patients who have undergone kidney damage due to diabetes or elevated blood pressure can lead to kidney failure over time. Two possible therapies for treating some of the symptoms associated with decreased kidney function include hemodialysis (HD) and peritoneal dialysis (PD). However, it is important to note that these therapies do not restore the proper functioning of the kidney. These procedures are commonly used as long-term treatments for patients with health issues associated with their kidney(s), and it provides these patients with an extended lifespan, which wasn't possible before these therapeutic interventions were designed.

Dialysis is the process of diffusing molecules in solution across a semipermeable membrane along an electrochemical concentration gradient (Depner, 1991). In patients with kidney damage, there is often an imbalance or disruption between the levels of intracellular and extracellular fluid. Therefore, HD aims to restore the balance between these fluid milieus to simulate normal kidney function (Himmelfarb & Alp Ikizler, 2010). In HD, the patient is given solutes transported through the blood. These solutes can move by diffusion or osmotic pressure gradients (Locatelli et al., 2002) through the tissues and organs of the body, restoring some levels of ion-balance homeostasis. PD is associated with renal complications including peritonitis (the inflammation of the inner lining of the abdomen), overhydration, electrolyte abnormalities, leakage and drainage difficulties of the renal system (Maxwell et al., 1959). There are different nutritional recommendations for patients with HD versus PD. Table 1* shows the recommended levels of electrolytes including sodium, potassium, phosphorus, calcium and fluid (Nanovic, 2005).

Table 1. Nutritional recommendations for hemodialysis and peritoneal dialysis patients

	Hemodialysis	Peritoneal dialysis
Sodium	<90 mEq Daily	Up to 150 mEq daily
Potassium	<60 mEq Daily	Up to 150 mEq daily
Phosphorus	800-1000 mg Daily	800-1000 mg Daily
Calcium	<2 g Daily	<2 g Daily
Fluid	1-1.5 L Daily	Up to 2 L daily

Source extracted from Nanovic, 2005

How do electrolytes link to renal abnormalities and COVID-19?

COVID-19 is a respiratory disease caused by the severe acute respiratory syndrome coronavirus type 2 (SARS-CoV-2) virus, originating from Wuhan, China in late December 2019. Patients were reported to have been experiencing contagious, pneumonia-like symptoms and these cases were linked to a variant of the SARS outbreak in 2002. There have been several cases of patients with COVID-19 and the onset of renal abnormalities. Further study into these cases has identified that electrolyte abnormalities are linked to kidney injury in COVID-19 patients. Kidney cells are partly responsible for facilitating viral entry, as they express the required receptors and enzymes (Farkash et al., 2020). Angiotensin-converting enzyme 2 (ACE2) is a protein associated with the entry of the SARS-CoV-2 virus. Many immunological conditions, such as hyperinflammation syndrome and cytokine storms, have shown to increase the onset of acute kidney injury (AKI) and glomerulopathy (González-Cuadrado et al., 1997; Sanz et al., 2011).

Further investigation of patients who developed severe COVID-19 symptoms have shown that previous renal conditions lead to more severe COVID-19 complications (Carriazo et al., 2020). In conditions such as hyponatraemia, there are issues associated with the secretion of antidiuretic hormone (ADH). ADH is a pituitary hormone, which signals for the kidney to release water and decrease the amount of urine produced.

There has been an increasing amount of discussion on current therapies for treating thrombosis (a blood clotting disorder) seen in patients who have contracted COVID-19. Current antithrombotic therapies work to decrease strong reactivity of the immune system in response to the presence of the SARS-CoV-2 virus in the body. Due to the nature of over-inflaming various regions within the body, this leads to blood clotting complications (Carriazo et al., 2020). As a result of the severe complications incurred by deep vein thrombotic events, it has been reported that the mortality rate of COVID-19 patients admitted to the ICU with thrombosis is at 50% (Nahum et al., 2020). It has been hypothesized that thrombosis in patients with COVID-19 can occur due to a variety of complications, including cytokine storms, hypoxic injuries, endothelial dysfunction, hypercoagulability, and increased platelet activity (Bilaloglu et al., 2020).

The role of exercise and physical activity
Exercise has profound effects on the overall health of the human body. Thermoregulation is responsible for regulating the loss of water and electrolytes. Studies have suggested that lower ion concentrations of sodium, potassium and chloride limit athletic performance and recovery (Coenen, 2005). Water balance and homeostasis is mediated through excretion of waste products, such as feces and urine. Electrolyte ion balances are maintained by the temperature-sensitive process of sweating in animals.

When prolonged exercise is sustained in warmer climates, this leads to a substantial decrease in water retention, therefore contributing to overall dehydration. In warm environments, there is an additional risk of developing heat-related illnesses, such as heat cramps, exhaustion and stroke. Electrolytes can be lost through water excretion in products such as urine and feces, as well as through organs including the lungs and skin (Maughan, 2003). It has been found that ionic sodium and chloride ions are lost primary through sweat during exercise or in high temperature environments (Costill et al., 1976).

Conclusion

In the human body, electrolytes facilitate a variety of roles for maintaining our overall human health. The upkeep and maintenance of the fluid-electrolyte balance with the human body has profound effects giving protection to obesity and cardiovascular disease risks. Electrolytes also mitigate the processes governed by aging, which control the homeostatic mechanisms linked to water absorption and excretion. As the renal system is primarily responsible for the regulation of electrolyte and ion balances within the body, it is important to keep these systems as healthy and strong as possible. Renal failure lowers the strength of our immune system, which is responsible for keeping us protected and safe from infectious diseases, such as COVID-19. When the renal system doesn't work well, our body is left to fight off excess toxins and waste products that have not been properly filtered out by the human body.

CHAPTER 3:

Role of Electrolytes in Human Health

Anna Yang

There are a number of different electrolytes in the human body, including sodium, potassium, calcium, chloride, magnesium, phosphate, and bicarbonate, and maintaining an appropriate balance of each of these electrolytes is vital for human health. Many crucial, automatic processes in the human body rely on small electric currents in order to function, and it is electrolytes that provide this charge as they interact both with each other and various cells. Electrolytes play key roles in processes such as nerve and muscle function, hydration of the body, and the maintenance of blood acidity. The principal functions of the principal electrolytes in the human body are outlined below.

Sodium
Sodium is the predominant extracellular cation and plays a key role in regulating the total amount of water in the body (Weiss-Guillet et al., 2003). Sodium's crucial role in maintaining water balance in the body arises from the fact that the amount of fluid in a body compartment depends on the concentration of electrolytes in it. If the concentration of electrolytes in a compartment is high, fluid will move into that compartment via osmosis. On the other hand, if the electrolyte concentration in a compartment is low, fluid will move out of that compartment. As such, the body is able to adjust the fluid levels in its various compartments by actively moving electrolytes in or out of cells through the use of various membrane proteins. This is key to proper health, as water must be kept in appropriate amounts both inside and outside of the cells in order for the body to maintain homeostasis.

In addition, the transmission of sodium into and out of cells facilitates critical body functions such as the generation of electrical signals in the nervous system and muscles (Stoppler, 2019). The characteristic electrical activity of these cells is the result of the opening and closing of specific ion-channel proteins in the plasma membrane, with the primary ions involved being sodium, potassium, chloride, and calcium (Lodish et al., 2000). These ion movements result in significant changes in the membrane potential and lead to the propagation of action potentials. At the resting potential, the voltage-gated ion channels are closed, and consequently there is no ion movement through them. However, when a region of the plasma membrane is slightly depolarized due to the binding of neurotransmitters to ligand-gated ion channels, voltage-gated sodium channels open for a short period of time, allowing for an influx of sodium ions (Lodish et al., 2000). This ion movement causes the sudden, though transient, depolarization associated with an action potential. Following the opening and closing of voltage-gated sodium channels, the transient opening of voltage-gated potassium channels causes the membrane potential to return to its resting state (Lodish et al., 2000). Thus, the ability of axons to conduct action potentials over long distances is dependent on the controlled opening and closing of voltage-gated sodium and potassium channels. Since the direction of ion movement through these channels also depends on the concentration gradient of these electrolytes across the cell membrane, electrolyte balance is crucial to these processes.

Calcium

Calcium is necessary for the effective functioning of many intracellular, cAMP-mediated messenger systems, most cell organelle functions, and many extracellular processes including muscle contraction, nerve signal transmission, and blood coagulation (Weiss-Guillet et al., 2003). For instance, striated muscles, which include cardiac and skeletal muscles, are regulated by calcium ions. Thin filaments in both skeletal and cardiac muscle cells contain tropomyosin and troponin, two proteins which form the troponin-tropomyosin complex (Sweeney & Hammers, 2018). In its resting state, the troponin-tropomyosin complex sterically blocks actin-myosin interactions in muscle cells. When calcium is released from the sarcoplasmic reticulum, however, it binds to troponin and induces a conformational change in the troponin-tropomyosin complex, allowing myosin to bind to actin (Sweeney & Hammers, 2018). This allows muscle cells, and therefore muscles, to contract.

Potassium

Potassium is predominantly an intracellular cation and plays a key role in both maintaining regular cell function and important bodily processes such as the regulation of heartbeat and the functioning of muscles (Kardalas et al., 2018). For instance, the membrane protein sodium-potassium ATPase, which pumps sodium out of cells and potassium into cells, is present in the membranes of nearly all cells in the human body and leads to the generation of a potassium gradient across cell membranes, making it partially responsible for the generation of a resting membrane potential (Kardalas et al., 2018). Many vital processes rely on this membrane potential, particularly in excitable tissues such as nerve and muscle cells, as outlined earlier. Enzyme activities, cell growth, and cell division are also catalyzed by potassium and are dependent on its concentrations, consequently making them sensitive to changes in its levels (Kardalas et al., 2018). Furthermore, intracellular potassium in particular participates in acid-base homeostasis throughout the body.

Magnesium
Magnesium, like potassium, is predominantly an intracellular cation. It is a cofactor in more than 300 enzyme systems which regulate diverse biochemical reactions such as protein synthesis, muscle and nerve function, blood pressure regulation, and blood glucose control (Ross et al., 2012). Magnesium is also necessary for functions such as energy production, glycolysis, and oxidative phosphorylation, and contributes to the structural development of bone and the synthesis of DNA and RNA (Ross et al., 2012). Finally, magnesium is involved in key interactions with other electrolytes, as it plays a role in the active transport of calcium and potassium ions across cell membranes, which are processes that are key to nerve impulse conduction and muscle contraction (Ross et al., 2012).

Phosphate
Phosphate is the most abundant intracellular anion in the human body and exists in numerous roles, including creatine phosphate, adenosine monophosphates and triphosphates, and more (Weiss-Guillet et al., 2003). Phosphate is essential for cellular structure and function, energy metabolism, bone mineralization, and genetic coding (Weiss-Guillet et al., 2003).

Given the vital roles played by electrolytes in various processes key to the maintenance of human health, it is hardly surprising that electrolyte imbalances can have detrimental impacts on people's health. Although patients are asymptomatic in many cases of electrolyte imbalances, they can also present with a wide variety of symptoms, and in certain cases, electrolyte imbalances can lead to medical emergencies. In fact, electrolyte disorders are among the most common clinical problems in intensive care units, with mixed electrolyte, acid-base, and fluid disorders being frequent occurrences as imbalances often occur in conjunction (Lee, 2010). The exact set of symptoms ultimately depends on the identities of the electrolytes that are out of balance and whether levels of the electrolytes are too high or too low. The clinical significance of electrolyte imbalances is well-established, with multiple studies reporting that electrolyte imbalances are associated with increased morbidity and mortality among critically ill patients (Lee, 2010). This chapter will discuss the health consequences of various electrolyte imbalances in order to highlight the key role played by electrolytes in the maintenance of proper health.

Hyponatraemia

Hyponatraemia is defined as a decreased concentration of serum sodium (<136 mmol/L) and is the most common electrolyte imbalance encountered in clinical practice (Upadhyay et al., 2006). The prevalence of hyponatraemia is particularly high among critically ill patients, with a study reporting the observation of hyponatraemia in 30% of a critically ill patient population, as their ability to excrete electrolyte-free water is often impaired (Upadhyay et al., 2006). It is worth noting that numerous studies have suggested that hyponatraemia in hospitalized patients is heavily underreported, with additional research being required in order to determine the true prevalence of this electrolyte imbalance in various clinical settings (Upadhyay et al., 2006). Hyponatraemia occurs when there is an increase in the amount of water relative to the amount of sodium in the body, which lowers the concentration of serum sodium, and can occur with numerous diseases including those of the liver and kidney (Stoppler, 2019). The development of hyponatraemia in critically ill patients in particular has been associated with disturbances in the renal mechanism of urinary dilution, which plays a key role in maintaining a proper balance of sodium and water in the body (Lee, 2010).

Patients with mild hyponatraemia (serum sodium concentration between 120 and 136 mmol/L) are generally asymptomatic, but lower values and rapid decreases can result in symptoms (Weiss-Guillet et al., 2003). The symptoms of hyponatraemia are primarily related to the central nervous system, with common symptoms including nausea and vomiting, headache, cognitive impairment, lethargy, restlessness, confusion, and in severe cases, seizures and coma (Weiss-Guillet et al., 2003). Seizures and coma typically occur only with rapid decreases in serum sodium concentration to levels lower than 125 mmol/L (Lee, 2010).

Hyponatraemia is not only highly prevalent among critically ill patients, but also has significant health consequences for this population. Both the disorder itself and treatments for the disorder have been associated with increased morbidity and mortality, with severe hyponatraemia in particular being associated with a high mortality rate among hospitalized patients (Upadhyay et al., 2006). While studies have demonstrated that it is likely the nature of underlying illnesses rather than the severity of hyponatraemia itself that best explains the mortality associated with the disorder, hyponatraemia nonetheless remains a strong predictor of mortality in the general population, independent of additional factors such as age, gender, and comorbidity (Chawla et al., 2011).

Hypernatraemia
Opposite to hyponatraemia is hypernatraemia, which occurs when there is an excess of sodium in the body relative to water. Patients in intensive care units are once again at high risk of developing hypernatraemia, with common causes being kidney disease, too little water intake, and loss of water due to diarrhea or vomiting (Stoppler, 2019). Symptoms generally do not manifest unless there is a sudden and significant increase in the serum sodium concentration to concentrations above 158 mmol/L (Weiss-Guillet et al., 2003). Non-specific symptoms such as anorexia, muscle weakness, restlessness, nausea, and vomiting tend to occur first, with more serious central nervous system dysfunction such as altered mental status, irritability, lethargy, or coma potentially following (Weiss-Guillet et al., 2003). All of these symptoms are the result of hypernatraemia's association with cellular dehydration and central nervous system damage, and are reflections of the key role that electrolytes such as sodium play in the maintenance of proper human health (Lee, 2010).

Hypocalcemia
Serum calcium levels are normally kept within a narrow range, between 2.1 and 2.6 mmol/L (Fong & Khan, 2012). When serum calcium levels drop below this range, hypocalcemia occurs and can range in severity from being asymptomatic to presenting as an acute, life-threatening emergency. The severity of symptoms is correlated with both the magnitude and rate of decrease in serum calcium concentration, and can also be influenced by additional factors such as acid-base status, magnesium levels, and sympathetic overactivity (Weiss-Guillet et al., 2003). Hypocalcemia is one of the most frequent electrolyte imbalances encountered in intensive care units, with studies reporting as many as

90% of critically ill patients being afflicted by low serum concentrations of calcium, with 15-20% having levels low enough to be classified as hypocalcemia (Lee, 2010). Similar to the aforementioned sodium imbalances, hypocalcemia has been associated with increased mortality among intensive care unit (ICU) patients (Lee, 2010).

With regards to these health concerns, acute hypocalcemia is the most relevant, as cases involving gradual decreases in serum calcium concentrations tend to be asymptomatic (Fong & Khan, 2012). Acute hypocalcemia, on the other hand, can result in several symptoms that require hospitalization. In cases of acute hypocalcemia, neuromuscular, neuropsychiatric, and cardiovascular symptoms tend to dominate (Weiss-Guillet et al., 2003). Common cardiac symptoms include bradycardia, ventricular dysrhythmias, and prolongation of the QT interval in electrocardiograms. Neuropsychiatric manifestations frequently include anxiety, irritability, depression, cognitive impairment, psychosis, and confusion, while neuromuscular symptoms include muscle spasms, cramps, and weakness (Weiss-Guillet et al., 2003). Paresthesia, tetany, circumoral numbness, seizures, laryngospasm, and dysphagia have also been observed (Fong & Khan, 2012).

Hypercalcemia
Hypercalcemia, elevated levels of serum calcium, develops when excesses of calcium cannot be excreted by the kidneys (Weiss-Guillet et al., 2003). The severity of the symptoms, as with hypocalcemia, is a function of the extent of the imbalance as well as the rate of onset, with mild hypercalcemia typically being asymptomatic (Weiss-Guillet et al., 2003). More severe cases of hypercalcemia, however, are associated with a broad range of non-specific symptoms including fatigue, weakness, anorexia, nausea, vomiting, depression, anxiety, vague abdominal pain, and constipation (Weiss-Guillet et al., 2003). Peptic ulcers have also been observed, likely due to calcium's role in stimulating gastric secretion, and particularly severe cases of hypercalcemia can result in acute pancreatitis (Weiss-Guillet et al., 2003). Cognitive dysfunction and personality changes are associated with serum calcium concentrations above 3.0 mmol/L while concentrations above 4.0 mmol/L can result in confusion, hallucinations, organic psychosis, somnolence, stupor, and coma (Weiss-Guillet et al., 2003).

Hypokalemia
Hypokalemia, characterized by serum potassium concentrations

28

below the normal range of 3.5-5.0 mmol/L, is a common electrolyte disturbance, particularly among hospitalized patients (Kardalas et al., 2018). It can arise due to a variety of causes, including kidney diseases, heavy sweating, endocrine factors, diarrhea, eating disorders, and certain medications (Stoppler, 2019). The commonality shared by these different causes is a root cause that relates to increased potassium excretion or intracellular shift, as hypokalemia rarely results from reduced potassium intake (Kardalas et al., 2018). Although hypokalemia sometimes requires urgent medical attention, severe hypokalemia (defined as potassium levels less than 2.5 mmol/L) are uncommon (Kardalas et al., 2018).

The severity of the symptoms of hypokalemia are generally proportionate to the extent of the potassium imbalance as well as the duration, and symptoms generally do not manifest until serum potassium concentrations drop below 3.0 mmol/L, with the exception of particularly rapid decreases (Kardalas et al., 2018). The symptoms of hypokalemia are highly varied, but can be roughly categorized according to the system affected. Symptoms related to the renal system include metabolic acidosis, rhabdomyolysis, and (rarely) impairment of tubular transport, chronic tubulointerstitial disease, and cyst formation (Kardalas et al., 2018). Effects related to the nervous system include leg cramps, weakness, paresis, and ascending paralysis (Kardalas et al., 2018). Symptoms related to the cardiovascular system include ECG changes (U waves, T wave flattening, and ST-segment changes), cardiac arrhythmias, and heart failure (Kardalas et al., 2018). In severe cases of hypokalemia, constipation, intestinal paralysis, and respiratory failure may be present (Kardalas et al., 2018).

Hyperkalemia
Hyperkalemia is characterized by increased serum potassium levels and can be caused by disorders that impair kidney function, as the kidneys are responsible for excreting potassium, as well as certain medications (Upadhyay et al., 2006). There is currently little agreement regarding classifications of mild, moderate, and severe hyperkalemia, but hyperkalemia in general is rare in normal patients because of a phenomenon known as potassium adaptation where the rate of potassium excretion in the urine increases following a high intake of potassium (Weiss-Guillet et al., 2003). As a result, the disorder is generally due to decreased potassium excretion as opposed to increased potassium intake.

Risk of hyperkalemia is particularly high among patients with severe infection or inflammation, direct muscle trauma, upper or lower motor neuron defects, renal failure, adrenal insufficiency, insulin deficiency or resistance, or burns, and as such the prevalence of hyperkalemia is highest among critically ill patients (Lee, 2010). This is exacerbated by the fact that many medications used in ICUs, such as beta-blockers, inhibitors of the renin-angiotensin-aldosterone system, potassium-sparing diuretics, heparin, and non-steroidal anti-inflammatory drugs, can also cause hyperkalemia (Lee, 2010).

Hypophosphatemia

Hypophosphatemia is characterized by low plasma phosphate concentrations, usually below 0.81 mmol/L, and has a number of potential causes including decreased intestinal phosphate absorption, increased renal phosphate losses, and shifts of phosphate to intracellular spaces. The prevalence of hypophosphatemia is particularly high in ICUs, with reports of it afflicting as many as 28% of critically ill patients. Symptomatic hypophosphatemia is generally observed in patients with a plasma phosphate concentration less than 0.32 mmol/L and the majority of the symptoms of hypophosphatemia can be explained by the resulting decrease in intracellular adenosine triphosphates which leads to failure of cell functions which are dependent on these energy-rich compounds. Noteworthy exceptions include osteopenia and osteomalacia, which are attributed to the effects of hypophosphatemia on mineral metabolism, and more specifically the direct effect of phosphate depletion on the osteoclastic resorption of bone. The clinical manifestations of hypophosphatemia are highly varied due to the large number of cellular mechanisms that are dependent on phosphate. Common symptoms include leukocyte, erythrocyte, and platelet dysfunction, muscular weakness, confusion, respiratory failure, cardiac arrhythmias, and neuropsychiatric disturbances.

Hypomagnesemia

Hypomagnesemia, abnormally low levels of serum magnesium, can be caused by a variety of factors including surgery, trauma, infection or sepsis, burns, excess gastrointestinal or renal losses, alcoholism, starvation, and certain medications such as diuretics, aminoglycosides, amphotericin B, cisplatin, digoxin, and cyclosporine (Lee, 2010). Like most electrolyte imbalances, hypomagnesemia is frequently observed among critically ill patients, with its prevalence in ICUs having reported to be as high as 50% (Lee, 2010).

The symptoms of hypomagnesemia overlap significantly with those of hypokalemia and hypocalcemia, with the primary clinical manifestations being neuromuscular and cardiovascular in nature (Weiss-Guillet et al., 2003). Additionally, due to the crucial role that magnesium plays in carbohydrate metabolism and insulin sensitivity, hypomagnesemia is associated with carbohydrate intolerance and hyperinsulinism (Weiss-Guillet et al., 2003). Particularly severe cases of hypomagnesemia can also result in electrocardiographic changes, seizures, arrhythmias, coma, and death (Lee, 2010).

Hypermagnesemia

Hypermagnesemia, elevated serum magnesium levels, is formally characterized by a serum magnesium concentration above 0.95 mmol/L and is rare, usually being the result of intravenous administration of magnesium or the use of antacids and laxatives containing magnesium (Weiss-Guillet, 2003). It is generally iatrogenic, which is to say that it is usually caused by a medical examination or treatment (Weiss-Guillet et al., 2003).

Common symptoms of hypermagnesemia include confusion, a depressed level of consciousness, muscular weakness, paralysis, vasodilation, hypotension, nausea, vomiting, and in severe cases, respiratory depression (Weiss-Guillet et al., 2003).

Conclusion

Electrolytes, through their interactions with each other and the cells in the body, facilitate a plethora of processes and functions that are essential to human health. As is often the case, the importance of the maintenance of proper electrolyte balance for healthy functioning becomes most apparent when this balance is disrupted, as electrolyte imbalances can result in a large variety of symptoms, ranging in severity from nausea and vomiting to coma and possibly death. Electrolyte imbalances are particularly common among critically ill patients, and this prevalence has a well-established clinical significance, with electrolyte imbalances being associated with both increased morbidity and mortality.

CHAPTER 4:

What are Electrolytes?

Amir Ala'a

A healthy body is what most and hopefully what all people strive to have. Eating healthy, and working out your body are widely known to be the stem of all good physical condition. Although maintaining a healthy lifestyle is indeed extremely beneficial, people tend to overlook the specific causes of their overall wellbeing. This can cause them not to achieve their fitness goals due to being unable to have possession of the knowledge regarding what their body specifically seeks out for. Electrolytes are one of those highly important things that most people overlook and fail to notice in order to help keep a healthy body. In order for us to understand how to properly benefit from electrolytes we must first fully understand what they are exactly and how they affect our well being.

There are seven main electrolytes that a human being is in need of to fit

the overall picture of health:
1. Sodium
2. Calcium
3. Chloride
4. Potassium
5. Magnesium
6. Phosphate
7. Bicarbonate

Sodium, potassium, calcium, and chloride are the most common electrolytes in the human body.

According to Shrimanker and Bhattarai (2020) one of the most essential electrolytes in the extracellular fluid is sodium, which is an osmotically active anion. It is in charge of regulating the extracellular fluid volume as well as cell membrane potential control.

As part of active transport, sodium and potassium are shared through cell membranes. The kidneys regulate sodium levels, and the bulk of sodium reabsorption takes place in the proximal tubule. Sodium is reabsorbed in the distal convoluted tubule. It is then transported by sodium-chloride symporters, which are regulated by the hormone aldosterone.

Potassium is primarily an ion found within cells. The sodium-potassium adenosine triphosphatase pump is in charge of maintaining sodium-potassium homeostasis by pumping sodium out of the cells in exchange for potassium entering the cells. The glomerulus filters potassium from the blood in the kidneys (Shrimanker & Bhattarai, 2020).

Shrimanker and Bhattarai (2020) state that Calcium plays a vital role in the body's physiological functions. It is involved in skeletal mineralization, muscle contraction, nerve impulse transmission, blood clotting, and hormone secretion. Calcium is primarily obtained via a person's. diet. It is primarily found in extracellular fluid. Calcium absorption in the intestine is mainly regulated by the hormones involved.

Chloride is an anion that is mostly present in extracellular fluid. The kidneys are primarily responsible for controlling serum chloride levels. The majority of the chloride filtered by the glomerulus is reabsorbed by both active and passive transport by both proximal and distal tubules (mostly proximal tubules) (Shrimanker & Bhattarai, 2020).

Phosphorus is a cation contained in extracellular fluid. The bones and teeth contain 85% of the total body phosphorus in the form of hydroxyapatite, while the soft tissues contain the remaining 15%. Phosphate is an essential component of metabolic pathways. It is found in a variety of metabolic intermediates, including adenosine triphosphates (ATPs) and nucleotides (Shrimanker & Bhattarai, 2020).

Magnesium is a cation that exists only inside cells. Magnesium is essential for ATP metabolism, muscle contraction and relaxation, neurological function, and neurotransmitter release. Magnesium causes calcium re-uptake by the calcium-activated ATPase of the sarcoplasmic reticulum as muscle contracts (Shrimanker & Bhattarai, 2020).

Bicarbonate levels are influenced by the blood's acid-base balance which the kidneys are in charge of maintaining the acid-base balance and regulating bicarbonate concentration. The filtered bicarbonate is reabsorbed by the kidneys, and new bicarbonate is generated by net acid excretion, which occurs when both titratable acid and ammonia are excreted. Bicarbonate loss is common in diarrhea (Shrimanker & Bhattarai, 2020).

Electrolytes are necessary to maintain a healthy and efficient body, they are one of the most required substances for human survival. Electrolytes are critical for many chemical processes in the body which rely on small electric currents to function, and electrolytes provide this charge. The functions in the body that require this current include muscle function (the movement of muscles when contacted) nervous system communication (the communication between the brain and nervous system), maintenance of body fluid levels, and regulating blood pH levels (Dr. Dave, 2018). The kidneys, along with many other hormones, regulate electrolyte levels. There are many traces of electrolytes in the human body such as the bloodstream, sweat and urine (Beswick, 2019). By definition electrolytes are a liquid or gel that contains ions and can be decomposed by electrolysis which is a "chemical decomposition produced by passing an electric current through a liquid or solution containing ions" (Oxford University Press, n.d.). A better definition of what they are is a substance that when dissolved in water will dissociate into positively and negatively charged ions that has the ability to conduct electricity in solution. Any substance that does not dissolve into ions when in water will not be able to conduct electricity, therefore called nonelectrolytes. The better the substance can break into ions while in water the stronger the electrolyte. Therefore, when looking at what is a strong electrolyte we must look at the efficiency of how many positively and negatively charged ions there are in solution (Anatomy and Physiology, 2019). To summarize the information stated, "Electrolytes are electrically charged minerals. In order for your cells, muscles, and organs to work properly, you need both fluids and electrolytes. Electrolytes help regulate the balance of fluid in the body" (Whelan, 2019).

To learn and understand ways we can incorporate healthy foods that contain electrolytes into our lives, we can refer to Canada's Food Guide. Canada's Food Guide (2021) recommends that the average plate of food for each person should consist of 50% fruits and vegetables, 25% whole grains, 25% protein packed food items (lean meats, nuts, beans, etc.), and of course we can not forget water on the side. Electrolytes such as sodium, chloride, potassium, and calcium are all we can obtain if we eat according to what the food guide recommends. Vegetables, as widely known, contain a great deal of nutrients, therefore, making them an extremely beneficial and trustworthy source to rely on in order to receive a healthy amount of electrolytes; this rewards them with the 50% of the plate they take up. Some vegetables high in electrolytes are spinach, kale, avocados, broccoli, potatoes, etc. Grains such as bread and pasta, on the other hand, tend to have a bad reputation in the dietary world due to containing carbohydrates which cause a scare in weight gain. That is a myth, carbohydrates do not cause weight gain but eating an excessive amount of them might, states Marisa Moore (2021).

Despite the stigma and criticism, grains have a lot of benefits in the electrolyte arena. As stated by Templeton (2020), grains bring in electrolytes like sodium which hold onto water causing you to stay hydrated for a long period of time. Protein dense foods hold magnesium which benefits bone development and your bones, being 60% of magnesium in your body, are clearly in desperate need to constantly contain that electrolyte (Whelan, 2019).

The food guide is not the only resource which gives you insight on how to manage your health. Multiple sport drink advertisements discreetly send information that benefit you and their product. For example, on July 30th 2020, Gatorade Canada posted an ad displaying men training and working out as a result of the energy Gatorade offers. If you listen closely, the ad begins by saying "With electrolytes to help replenish what's lost in sweat," and they are correct. When it comes to working out, an athlete's body sweats profusely causing electrolytes to be drawn out of their system, leaving the brand with the perfect opportunity to manufacture and deliver a drink that contains sodium and potassium to "help fuel your working muscles" during your active and overall regular state. Unfortunately, sport drinks are also packed with added sugars and sweeteners making them not the healthiest but for sure one of the most convenient ways to consume your electrolytes. Not only can you eat and drink electrolytes but you can also gain your vitamins and minerals through supplements. Due to the fact that electrolytes are lost during sweat and exercise, not having those minerals in the body to support workouts can be dangerous. It is also a preference to not be consuming high amounts of sugars (that energy drinks may offer) as a result of diets, health conditions, and other lifestyles, and electrolyte tablets (that come in different types of electrolytes which can also help with imbalances) come in incredible convenience. Although if you have any pre-existing health concerns and/or you are already taking medication, it is best to consult your doctor before moving any further with electrolyte pills or capsules. Taking large quantities of electrolytes without consulting a doctor can cause another imbalance and result in serious health complications. In certain cases, further care needs to be taken to straighten out the underlying problem. Electrolyte imbalances can be dangerous. If the imbalance is severe, electrolytes can always be administered through the use of an intravenous, or IV, line in a hospital (Berry, 2020).

"Fluids are found inside and outside the cells of your body. The levels of these fluids should be fairly consistent. On average, about 40 percent of your body weight is from fluids inside the cells and 20 percent of your body weight is from fluids outside the cells. Electrolytes help your body juggle these values in order to maintain a healthy balance inside and outside your cells" (Whelan, 2019). When the amount of water in your body increases, electrolyte levels may become too low or too high. The amount of water you consume should be equal to the amount you waste. You could have too little water (dehydration) or too much water (overhydration) (U.S. National Library of Medicine, 2021). Electrolytes are so incredibly important that when their levels in the body are either too high or too low they can manage to introduce electrolyte disorders. An extended amount of diarrhea, vomiting, or sweating are most often the origin of electrolyte disorders. Depending on the particular imbalance, the actual cause as to why it developed may vary. The prefix "hyper" is used to indicate that there is an extensive amount of an electrolyte in the body, but when it is the opposite, the prefix "hypo" is used. For example calcium disorders are referred to as hypercalcemia (too much calcium in the bloodstream) and hypocalcemia (lack of calcium in the bloodstream). These disorders can also be created by other serious issues one can be experiencing such as:
• Severe dehydration; a medical emergency that could be life-threatening. It has the potential to damage your kidneys, heart, and brain.
• Eating disorders; any of a number of psychiatric conditions marked by irregular or erratic eating patterns
• Severe burns; extreme skin damage that results in the death of the affected skin cells
• Severe breathing difficulties; an intense tightening in the chest
• Heart failure
• Alcohol use disorder
• Kidney failure, and many more (Holland, 2019).

There can also be recurring events where there is a lack of electrolytes which is not as intense as a disorder, and they can appear because of common and natural electrolyte fluctuation. The human body cannot function properly unless there is a normal electrolyte balance, some symptoms of an imbalance include: nausea, lethargy, fatigue, irregular heart rate, muscle cramps, confusion, constipation, and more. Eating food and drinking fluids rich in electrolytes can help replenish the lack of nutrients. Eating healthy and being hydrated with the right amount of water each day (eight, 8 oz glasses a day) positively affects your body and its electrolytes immensely (Buchanan, 2020). Staying in balance with your electrolytes is something people do subconsciously on a day-to-day basis. Some easy, free, and effective ways to maintain your electrolyte balance are maintaining a healthy diet, drinking water but not overdoing it (too much can flush out electrolytes from the body), not overusing salt (despite it being an electrolyte bodies need, too much of it can throw off your electrolyte balance), etc. It can also come in huge assistance to visit a doctor or nutritionist in order to come to a conclusion on figuring out the right electrolytes your own unique body craves.

Because of their smaller size and faster fluid electrolyte metabolism, children tend to be more prone to dehydration than adults. If a child falls ill with heavy diarrhea or vomiting, an electrolyte imbalance may develop, which will need medical attention. An electrolyte imbalance in a child with an underlying health condition, such as thyroid, heart, or kidney disease, is more probable. Dehydration and electrolyte imbalances could be more common in older adults than that in younger adults, according to studies (El-Sharkawy et al., 2013). This is due to a variety of factors, including:
Kidneys losing some of their functions with age
Multiple medications, such as diuretics, can be consumed among elderly individuals, causing electrolyte levels to fluctuate
Caregivers and loved ones should maintain an eye out for signs of dehydration within the elderly. They will also need to aid in guaranteeing that the patient is receiving enough meals and drinks (Berry, 2020).

In quick summary, a healthy body is what each and every person aims to achieve, regardless of age, religion or gender. In order to optimize the ability of our body's long-term function, we must be able to understand electrolytes, their benefits, health risks, and science, in order to intake the proper nutrition they deliver, and to successfully move past our day-to-day activities in good shape. Many people tend to look over and disregard electrolytes, despite the fact that they are in control of a large number of critical operations by the human body. Electrolytes are crucial for an obscene amount of chemical processes that take place in the body which rely on miniscule electric currents in order to carry out their duties. Electrolytes provide this charge, which without a doubt proves just how much our bodies depend on their effects. Canada's Food Guide, energy drink/bar/capsule/pill companies end up developing and recommending items and lifestyles whose main star of the show are electrolytes. This then opens up an opportunity for brands to make use of the electrolyte's undeniably large importance and create and advertise products they can financially benefit off of, and for us to physically take full advantage of. Because of their enormous value, an imbalance of electrolytes can cause many serious health issues if not addressed and taken care of immediately. If there truly is a severe deficiency, then a hospital visit is a necessary action to regain your body's electrolytes through an IV, etc. Older adults tend to be at more of a risk of ending up with an electrolyte disorder due to the fact that as someone ages, their health diminishes along with them. The same applies with young, continuously developing children, in the sense that their bodies are still gaining familiarization with illnesses.

CHAPTER 5

The Positives and Drawbacks of Popular Electrolyte Drinks

Angel Xing

Introduction to Popular Electrolyte Drinks

Electrolytes, as essential minerals for the body's chemical functions, are fundamental for human health. However, humans continuously lose electrolytes throughout the day, primarily depleted through sweat and urine. While water and regular meal intakes can replenish lost electrolytes, these methods can be insufficient for many individuals. Anyone who excessively loses fluids would also lose electrolytes, which are important minerals like potassium, calcium, and chloride. Key examples of such individuals include athletes who would generate more sweat, those suffering from heatstroke, and those undergoing gastrointestinal issues including, vomiting and diarrhea. In short, any activity resulting in dehydration, the excessive loss of fluids, will also result in the loss of electrolytes. Unfortunately, water or food may be insufficient to replenish these high levels of dehydration. Fortunately, there are various electrolyte drinks designed for this purpose. This chapter will explore the positives and drawbacks of two main types of popular electrolyte drinks: electrolyte-infused beverages including sports drinks and electrolyte-infused water, and oral rehydration solutions like Pedialyte.

One primary type of electrolyte drink is electrolyte-infused drinks. These include sports drinks and electrolyte-infused water. Commonly sold under popular brands like Gatorade and Powerade, sports drinks are readily accessible across Canada. Not to be confused with energy drinks like Red Bull, sports drinks are designed to replenish glucose, fluids, and electrolytes lost by athletes during exercise (Pound et al., 2017). Unlike energy drinks, sports drinks do not contain any caffeine. According to data provided on the U.S Department of Agriculture's FoodData Central, Gatorade Lime's ingredients include water, sugar, dextrose, citric acid, salt, sodium citrate, monopotassium phosphate, modified food starch, natural flavour, and food colouring (Food Data Central, n.d). Evident by the presence of sugar, modified food starch, sodium citrate, and monopotassium phosphate, the ingredients for sports drinks are chosen for quick hydration and absorption. While scientists found them to be effective for athletes, it was more difficult to determine the correlation between the drink's efficacy and exercise time for children as they sweat at variable rates (Pound et al., 2017). Regardless, sports drinks are a popular consumer item and a key example of an electrolyte drink.

Electrolyte-infused water is another form of electrolyte-infused drink. Brands like Essentia Water sell alkaline water infused with electrolytes. First, their water undergoes an ionization process to have a higher pH, thus becoming alkaline, and then they add electrolytes like sodium bicarbonate, dipotassium phosphate, magnesium sulphate, and calcium chloride (Prescribers' Digital Reference, n.d). While both sports drinks and electrolyte-infused water includes electrolytes, their purposes are different. First, electrolyte-infused water does not contain sugar because it does not aim to replenish glucose. Second, sports drinks are designed to restore electrolyte levels, but electrolyte-infused waters are not marketed as such. Instead, Essentia markets its product as a "supercharged ionized alkaline water" (Essentia, n.d). Nevertheless, electrolyte-infused waters are another example of drinks marketed to boost electrolyte levels.

The second main type of electrolyte drink is oral rehydration solutions (ORS), like Pedialyte. Although they are the same as electrolyte-infused beverages in that they contain electrolytes, their purpose is different. ORS or oral rehydration treatment (ORT), are primarily used to treat excessive vomiting and diarrhea, mainly in infants and children. Since electrolytes can be lost through any loss of bodily fluids, diarrhea and vomiting episodes would consequently result in excessive losses of electrolytes. Thus, ORS and ORT would replenish these electrolytes. In the fruit-flavoured Pedialyte, the medicinal ingredients include dextrose, citric acid, potassium citrate, sodium chloride, sodium citrate, and zinc gluconate. The non-medicinal ingredients include water, natural and artificial flavours, sucralose, acesulfame potassium, and food colouring (Pedialyte, n.d). Given that Pedialyte is mainly used as a treatment, it is often forgotten that it is also a popular type of electrolyte drink.

There are also other natural drinks with electrolytes, most popularly milk and coconut water. However, this chapter will focus on comparing the positives and drawbacks of the three main artificially created electrolyte drinks: sports drinks, electrolyte-infused water, and ORS and ORT like Pedialyte.

Positives of Electrolyte Drinks

Part 1: Electrolyte-infused beverages

Electrolyte-infused beverages have many positive aspects, especially as sports drinks are marketed towards athletes. Since athletes lose fluid during physical activity, they would subsequently lose glucose and electrolytes (Sawka et al., 2007). Therefore, researchers recommend for athletes to aim to drink an electrolyte beverage, such as a sports drink, before exercise. This way, they can better prevent excessive dehydration, which is considered to be a greater than 2% body weight loss from water deficit. The investigators acknowledge that there are many variable factors to how much sweat an athlete can generate, thereby causing variable rates in the loss of electrolytes. Such factors include body weight, genetic predisposition, heat acclimatization state, metabolic efficiency, and the environment as a warmer environment typically results in more sweat generated. Thus, trainers often monitor their athletes' electrolyte and fluid loss levels to decide on the optimal recovery paths (Baker, 2017). By using sweat testing, trainers can determine the best electrolyte replacement strategies (Baker, 2017). Since scientists found that different athletes have different needs, they concluded that sports drinks are beneficial as an electrolyte replacement, prevention of dehydration, and useful for pre-exercise and post-exercise hydration, but individual variability must also be taken into consideration (Orru et al., 2018). Since sports drinks also contain glucose, they would also replenish energy. On the other hand, electrolyte-infused water does not typically contain glucose, thus it would not necessarily be the best option for athletes.

Nonetheless, all muscle contractions would produce metabolic heat, elevating the body's core temperature. As a result, the body undergoes physiological adjustments to cool itself by generating sweat (Sawka et al., 2007). Consequently, thermoregulatory sweating during exercise would result in water and electrolyte losses (Baker, 2017). Although athletes can wear permeable material to reduce evaporative cooling requirements, they will still irrefutably lose electrolytes through sweat (Sawka et al., 2007). Thus, it is imperative to replenish water and electrolyte losses to prevent dehydration.

Moreover, the intake of sports drinks before exercising would also prevent excessive changes in electrolyte balance (Sawka et al., 2007). During an athlete's routine, researchers said that drinking electrolyte beverages would be more beneficial than water, and that post-exercise, it is imperative for athletes to replace their fluid electrolyte deficit (Sawka et al., 2007). However, scientists believe that prevention is more crucial than recovery, therefore they advise athletes to consume sports drinks before exercising (Sawka et al., 2007).

Sports drinks are necessary for peak athletic performance. If athletes did not replenish their electrolyte and water levels after exercising, this deficit will result in an increase in cardiovascular and thermal strain, which would negatively impact aerobic performance (Shireffs & Sawka, 2011). While some athletes can recover their electrolyte and fluid levels by normally eating and drinking after practices to re-establish euhydration, the normal total body water level, this practice does not apply to all. Athletes who must recover under 24 hours either due to training regimes or competitions, as well as those facing severe hypohydration where greater than 5% of their body mass of water is lost, should consume sports drinks (Shireffs & Sawka, 2011). Indeed, sports drinks containing sodium can improve athletic performance by sustaining the athlete's metabolism and aiding in the optimization of water absorption (Orru et al., 2018).

In addition, electrolyte beverages, like sports drinks, are also beneficial for muscle cramps (Lau et al., 2021). Exercise-associated muscle cramps (EAMC) are caused predominantly by dehydration and electrolyte deficits. While it appears that drinking water would solve the issue, researchers found that drinking a large volume of plain water would dilute sodium and electrolyte levels in the athlete's blood and extracellular fluid, which would increase EAMC susceptibility (Lau et al., 2021). Ultimately, scientists concluded that spring water intake during exercise increase muscle cramp susceptibility while the intake of OS-1, an electrolyte beverage made by Otsuka Pharmaceutical Factory containing sodium, potassium, magnesium, chloride, glucose, and phosphorous, resulted in the decrease of muscle cramp susceptibility (Lau et al., 2021). In short, electrolyte beverages, like sports drinks, can decrease muscle cramps, which is an additional positive for athletes.

Another positive aspect of sports drinks and electrolyte-infused water can be their potential benefit for the management of heat stroke and heat exhaustion (Armstrong, 1994). Researchers studied the effect of carbohydrate-electrolyte replacement fluids on soldiers in various military field situations, concluding that these fluids were most beneficial for soldiers under specific conditions. Those who lost more than eight litres of sweat per day, were not heat acclimatized, performed prolonged and continuous exercise greater than 60 minutes, had inconsistent meals, experienced a caloric deficit of greater than 1,000 kcal per day, and/or had diarrhea were in greatest need of the fluids, and had the greatest response. Indeed, one great factor affecting the soldiers' needs for the carbohydrate-electrolyte replacement fluids was the effect of heat.

Heat-related illnesses are caused by inflammatory pathologic events that start with mild heat exhaustion, and if left untreated, can lead to multiorgan failure and death (Glazer, 2005). Common symptoms of heat-related illnesses include headache and nausea, and typical treatment methods are focused on cooling the patient through relocation to a different environment and hydration. There are four main ways of how heat is exchanged with the environment: conduction, convection, radiation, and evaporation. Conduction is the loss of heat through direct contact with a cooler item, convection is the dissipation of heat, radiation is the release of heat, and evaporation is achieved through perspiration where the body generates sweat to cool the body. Given that evaporation is the most effective method for the body to cool itself, dissipating up to 600kcal per hour in optimal conditions, it is also the main method. Thus, when the body is in high heat situations, it will perspire to cool down. Unfortunately, as established earlier in this chapter, excessive sweating results in dehydration and the loss of electrolytes. In the attempt to cool down, the body will become more dehydrated, thereby requiring immediate rehydration (Glazer, 2005). Therefore, sports drinks and electrolyte-infused water can replenish electrolyte levels. However, they are not a treatment for heat stroke and heat exhaustion, merely another fluid replenishing alternative.

Part 2: ORS and ORT

ORS and ORT, like Pedialyte, have many positive aspects. Since they are designed to rehydrate children with gastrointestinal issues, like vomiting or diarrhea, they are especially beneficial in the treatment and management of acute gastroenteritis (Nazarian, 1997). Scientists recommend for ORS and ORT to be used to replenish lost fluids and electrolytes caused by diarrhea in young children, aged one month to five years when the child is moderately dehydrated. Through diarrhea, water, sodium, potassium, and chloride would be depleted from the body.

In the 1930s, doctors used intravenous (IV) therapy as the first successful routine method to administer fluids and electrolytes (Nazarian, 1997). This was widely accepted as the standard form of rehydration therapy. However, this practice was more difficult to conduct on children, so they developed ORT. Not only was ORT successful, proven by rehydrating more than 90% of the studied dehydrated children suffering from cholera, it also had fewer complications than IV therapy. Additionally, it was substantially less expensive than IV therapy. However, scientists did note that glucose-electrolyte ORT had no effect on stool volume or the duration of diarrhea, and so it must be used in combination with another treatment. While it can replenish lost fluids and electrolytes, it would not treat the root of the gastrointestinal issues (Nazarian, 1997). Nevertheless, ORT and ORS have significant positive impacts on the treatment of children's gastrointestinal issues.

Drawbacks of Electrolyte Drinks

Despite the positive aspects of popular electrolyte drinks, there are also key drawbacks to consider. This section will explore three primary issues: the high sugar content of sports drinks, the high content of salt in electrolyte beverages, and the side effects of Pedialyte.

Part 1: High Sugar Content

Sports drinks are typically considered functional beverages because they are, as defined by the European Commission, a beverage that has "beneficial effects on one or more function(s) of the human organism, thus improving the general physical conditions or/and decreasing the risk of the evolution of disease" (Orru et al., 2018). However, scientists warn that "intense marketing efforts are continually made to encourage the consumption of functional beverages, even when they are not needed" (Orru et al., 2018). That is to say, sports drinks can be considered beneficial based on the virtue of their naming and advertising, but they may be mass-marketed beyond the intended consumers.

While sports drinks include electrolytes, they also include a large mass of sugar. Based on information from the U.S Department of Agriculture's Food Data Central, 355 millilitres of lime-flavoured Gatorade includes 20.4 grams of sugar (Food Data Central, n.d). Another popular sports drink, lemon-lime-flavoured Powerade, includes 22.3 grams of sugar for 365 millilitres of the drink. In the entire drink, carbohydrates make up 115 out of 117 total calories. Comparatively, for every 360 millilitres of Pepsi in a 12 fluid ounce glass bottle, there are 39 grams of sugar. In one Coca-Cola 12 fluid ounce glass bottle of 360 millilitres, there are also 39 grams of sugar. While Gatorade and Powerade still contain less sugar than Pepsi and Coca-Cola, it still contains high levels of sugar as the American Heart Association (AHA) recommends for a maximum of 24 grams of added sugar per day for women, and under 36 grams of added sugar for men (Johnson et al., 2009). The investigators cited causes for the rising consumption of sugar in the average American's diet predominantly due to soft drinks and other sugar-sweetened beverages, leading to metabolic abnormalities and adverse health conditions. Indeed, popular sports drinks include electrolytes, but they also include high levels of sugar.

Moreover, scientists found an association between sports drinks with weight gain among adolescents and young adults (Field et al., 2014) Through following 4,121 females and 3,438 males aged 9-16 from the United States between 2004 to 2011, researchers concluded that the intake of sports drinks predicted larger increases in BMI among both females and males. Since sports drinks are portrayed as part of an active lifestyle, the researchers noted that there were few studies on the actual relationship between sports drinks and weight gain. In addition, scientists noted that sports drinks were typically sold in 20 ounces, around 567 grams, bottles. Thus youth would automatically consume more sugar. Ultimately, the study observed a positive association between the intake of sports drinks and weight gain as it still has high sugar content, which can increase risks of obesity and type II diabetes (Field et al., 2014). Evidently, the high sugar content in sports drinks is a drawback.

Additionally, sports drinks are a commonly consumed sugar-sweetened beverage by adolescents. In a 2012 study of 11,029 adolescents, the investigators found that 26% of sugar-sweetened beverages were sports drinks while 37% were regular soda (Park et al., 2012). During Field's (2014) investigation, the team observed that nationally, consumption of regular soda decreased between 2004 and 2011, which was mirrored by the youth study. They hypothesized that sports drink endorsements by sports celebrities could have subsequently resulted in an increased interest in sports drinks and a decreased interest in soda (Field et al., 2014). However, sports drinks were designed for athletes, especially for those participating in rigorous or endurance training lasting longer than 60 minutes. As well, the study cites that there are conflicting studies about the actual biological impacts of sports drinks on athletes. The study also indicates that there was no research on the efficacy of sports drinks for short-term activity and the prevention of dehydration for non-athletes (Field et al, 2014). Therefore, sports drinks are also marketed and consequently consumed by non-intended audiences who experience weight gain from the high sugar content.

The artificial sweeteners in sports drinks are also detrimental to the gastrointestinal system. FDA-approved examples of artificial sweeteners include acesulfame, aspartame, neotame, saccharin, and sucralose (Spencer et al., 2016). Especially for patients with irritable bowel syndrome (IBS), many artificial sweeteners would trigger gastrointestinal symptoms. While artificial sweeteners are often used to decrease caloric content as they contain fewer calories than sugar, the review found that artificial sweeteners affect two main areas: motility and the gut microbiome. They can cause alterations in gastrointestinal motility by affecting the secretion of incretin hormones, delaying gastric emptying, intestinal transit, and increasing serotonin which initiates peristalsis. As well, various studies examined stool samples of animals and humans, finding that artificial sweeteners affect the gut microbiomes. Since Gatorade includes dextrose, an artificial sweetener derived from corn syrup, as its third ingredient, it may be detrimental to the gastrointestinal system. Therefore, the inclusion of artificial sweeteners in sports drinks can be irritating for individuals with IBS (Hayes et al., 2014).

Part 2: High Sodium Content

Electrolyte drinks also often include a high level of sodium. Consequently, this can result in higher blood pressure (Iqbal et al, 2019). By lowering sodium and increasing potassium intake, researchers found that blood pressure could be lowered, especially in hypertensive patients (Iqbal et al, 2019). Although sodium and potassium are both considered electrolytes, they are still different elements with separate properties. Since Pedialyte's medicinal ingredients include potassium citrate, sodium chloride, and sodium citrate, it would raise both sodium and potassium levels. High levels of sodium are associated with morbidity and mortality from cardiovascular diseases. High sodium intake often results in water retention, leading to increased blood pressure levels (Grillo et al., 2019). In addition, high sodium levels would alter endothelial function, change the structure and function of large elastic arteries, modify sympathetic activity, and affect the autonomic neuronal modulation of the cardiovascular system (Grillo et al, 2019). From this study, scientists concluded that the general population should reduce their salt intake as the worldwide average ranged between 3.5-5.5 grams when the World Health Organization recommends a limitation of 2.0 grams (Grillo et al, 2019). Therefore, the excess of sodium in electrolyte-infused drinks is a major drawback as it can raise sodium levels.

Sodium in electrolyte-infused beverages, like sports drinks, may not be beneficial to athletes either. Through a study on the effect of sodium supplements and climate on dysnatremia, abnormal sodium concentration, during ultramarathon running, the investigators found that sodium supplementation did not prevent or cause exercise-associated hyponatremia, low sodium concentration in the blood (Lipman et al, 2020). Instead, the most important factors to prevent low sodium concentration in the blood were longer training distances, lower body mass, and avoiding overhydration (Lipman et al, 2020). However, sodium supplements also did not cause hypernatremia, excess sodium in the blood (Lipman et al, 2020). As well, this was only the first study. There warrants more investigation into the impact of electrolyte-infused sports drinks on the body at a deeper level.

Part 3: Pedialyte Side Effects

Even though Pedialyte effectively replenishes fluids and electrolytes after they are lost through vomiting and diarrhea, it, like all medications, has side effects. Common ones include nausea, vomiting, diarrhea, and stomach pain, which can be alleviated by mixing the medication with water or juice, or by taking it after meals (RxList, 2019). As well, the high sodium concentration may negatively affect high blood pressure during pregnancy (RxList, 2019). While it appears that Pedialyte, as an electrolyte-containing drink, would be beneficial for quenching dehydration, it is key to remember that Pedialyte is a medication. Like all medication, Pedialyte is advised to be taken based on the appropriate dosage.

Conclusion

All in all, there are several types of popular electrolyte drinks. This chapter focused on two main categories: electrolyte-infused beverages, like sports drinks and electrolyte-infused water, and medication (ORS and ORT), like Pedialyte. Although there are several positives to electrolyte-infused beverages, such as the replenishing of electrolytes lost through vomiting, diarrhea, or sweat, there is additional research necessary to examine how sports drinks affect the body on a cellular level, and if it has enough positive impacts to match the marketing depiction. Moreover, while sports drinks appear to benefit athletes, it also presents a key drawback: the high sugar concentration. While sports drinks suit a select population, athletes who undergo strenuous and long intervals of training, it is marketed towards the general public, appealing with its sweet taste but seemingly low sugar concentration. While the sugar content is lower in sports drinks compared to soda, scientists found that it does negatively affect consumers who do not regularly exercise and practice athletic training. Thus, electrolyte-infused beverages have both positives and drawbacks to consider.

ORS and ORT, like Pedialyte, are another popular form of electrolyte drinks. They have many positives, for they are the preferred treatment over IV therapy for gastrointestinal issues in children. They are also more easily applied and accessible than IV therapy. However, there are also important side effects to observe. Additionally, ORS and ORT are fundamentally medications, so they cannot be consumed without consulting the recommended dosages. Both types of electrolyte drinks include high levels of sodium, and excessive sodium intake would have a negative impact on the human body. It is necessary to follow recommended health guidelines and ensure a balanced level of electrolytes in the body. When electrolytes are lost, they should be replenished. When they are not lost, then there is no need to have an excess.

CHAPTER 6

The Science Behind How Electrolytes Work

Amal Rizvi

Electrolytes naturally occur in the body, and thus, electrolyte balance is in tandem with nutrient balance and fluid balance; electrolyte, nutrient, and fluid balance are consequently a function of bodily processes such as metabolism, absorption, digestion, and ingestion (Allison, 2004). They are essentially charged minerals, known as ions, that are balanced with positively changed minerals, such as, but not limited to sodium, magnesium, or potassium, and a negatively charged mineral complex, such as chloride, sulphate, or carbonate (Elete Electrolyte, n.d.). It is important to note that the word "electrolyte" is often used as a blanket term that encapsulates minerals that can also be found in a non-electrolytic form. For example, magnesium chelate is called an electrolyte, but it does not function as an electrolyte. This is because electrolytes function by dissolving and splitting in water, in order to make fluid conduct electrical impulses. Magnesium chelate will not behave in this manner, because it does not contain a negatively charged mineral complex. The electrolyte magnesium chloride, on the other hand, behaves like a perfect electrolyte, because chloride is a negatively charged mineral complex that allows the ion to dissolve in water.

When it comes to beverages, electrolytes are present in sports drinks, or isotonic drinks (RSC, 2020). These beverages contain similar concentrations of salts and glucose as in the cells of the body. Their purpose is to make up for the electrolytes athletes lose during or after exercise. It is important to note that these types of beverages are not any more hydrating than plain water is, but athletes typically drink much larger volumes of sports drinks that they would drink of water in order to replenish their lost electrolytes. They are lower in calories than soft drinks are, but they can contain supplemental carbohydrates, thus also providing a carbohydrate boost to the athlete.

It is important to also keep in mind the difference between sports drinks and energy drinks. Energy drinks are beverages that contain caffeine, glucose, as well as taurine, guarana, and ginseng, allegedly for the purpose of boosting an individual's performance and stamina (RSC, 2020).

In the human body, electrolytes are present in both the extracellular fluid - the fluid outside of the cells, including plasma, lymph, cerebrospinal fluid, synovial fluid, and interstitial fluid - and the intracellular fluid, which is the fluid inside of the body's cells (MRI, 2007). Dilute solutions of naturally occurring electrolytes, such as sodium, chloride, and bicarbonate exist within the intracellular and extracellular fluid. These electrolytes are then used by the organs and tissues of the body to carry electrical signals between cells, as well as for the purpose of nervous system and muscle regulation.

Additionally, the kidneys are a fundamental regulatory organ in regards to electrolyte balance. This is because the kidneys maintain fluid absorption and secretion, which enables them to regulate electrolyte levels in the human body (MRI, 2007). They can fitler electrolytes back into the circulation for the purpose of the body utilizing them for a specific function. Conversely, if the level of a particular electrolyte, such as sodium or potassium, is higher than necessary at a given moment, the kidneys facilitate the excretion of that electrolyte through the urine or feces.

When an individual exercises, their fluid and electrolyte balance shifts; this is the consequence of water and electrolyte loss with sweat (Brouns & Kovac., 1997). Essentially, during exercise, arterial pressure increases, which increases capillary hydrostatic pressure (Conventino, 1987). This increase in capillary hydrostatic pressure favours a shift in plasma volume from the vascular compartment to the interstitial fluid, which results in more fluid available overall for sweat production (Hinchcliff et al., 2008). The effects of electrolyte loss from sweating typically decrease within a few hours after exercising, after which the urinary excretion of electrolytes - primarily magnesium and potassium - increase. In addition to sweating, electrolyte levels can change as a result of rapid fluid loss; the classic example of this is after a rapid bout of vomiting or diarrhea (Felman, 2017).

Drinking alcohol can also affect electrolyte levels. This is because alcohol is a diuretic, which, in simple terms, means that it results in an individual excreting more fluids, and thus electrolytes, through urination (Beswick, 2019). Alcohol essentially inhibits antidiuretic hormone (ADH), which is the hormone that helps control how the human body retains water and electrolytes, rather than promoting their loss through urine. When ADH is suppressed, an individual urinates more, thus losing more electrolytes. Further, when individuals are consuming a lot of alcohol, they are not usually re-hydrating themselves with enough water (Beswick, 2019). As a consequence of this, individuals consuming large volumes of alcohol are going to end up losing even more electrolytes, especially if they vomit by the end of the night.

Sodium, chloride, and potassium are the main electrolytes that are lost through sweating (Orrù et al., 2018). This can be harmful to individuals during exercise, as electrolytes are crucial for various biological functions in the body. For example, potassium and sodium are fundamental in regulating body water via sodium reabsorption and potassium secretion in the distal convoluted tubule and the connecting tubule in the kidneys; this mechanisms allows the body to maintain homeostasis (Elete Electrolyte, n.d.). Potassium is the more abundant electrolyte inside cells, and is also necessary for muscle function, nerve conduction, and pH balance, whereas sodium is the least energetically expensive electrolyte, and the most common electrolyte found in individuals' diets. (Elete Electrolyte, n.d.). Sodium is also necessary to maintain a functioning thirst response, heat tolerance, nerve condition, and pH balance. Chloride is another important electrolyte, as it is responsible for maintenance of acid-base balance, osmotic pressure, and comprising a significant portion of gastric juice (Orrù et al., 2018). Chloride is the most abundant negatively charged electrolyte in the human body, and it also has a crucial role in oxygen exchange (Elete Electrolyte, n.d.). Magnesium is also a fundamental electrolyte, and is actually the most expensive electrolyte, in terms of energy consumption. Magnesium is crucial because it is the basis for all energy conversion in the body. Further, it aids in muscle functioning, the nervous system, enzymatic biochemical reactions, and bone formation. The main way magnesium is depleted from the body is through physical and emotional stressors (Elete Electrolyte, n.d.).

Electrolyte loss during sweating is a major reason why athletes and individuals who partake in exercise consume beverages with added electrolytes. However, generally, sodium is the only electrolyte that needs to be added to these beverages - this is typically in the form of sodium chloride (NaCl) or other salts, like sodium citrate (Shirreffs, 2009). When sodium is consumed in the form of an electrolyte-enhanced beverage, it stimulates the uptake of glucose and water in the jejunum of the small intestine. This, consequently, aids the body in maintaining high plasma osmolarity as well as the maintenance of the extracellular fluid. However, although high sodium in electrolyte-enhanced beverages can provide benefit through increased jejunal uptake of glucose and water, it is important to keep in mind that it can make drinks unpalatable. Another important consideration is that electrolyte replacement is not necessary for most athletes and/or individuals who partake in exercise, unless they are participating in exercise or sport for a very long period of time, in which the body is expected to expel large volumes of sweat (Shirreffs, 2009).

Electrolyte imbalance is a condition that occurs when either electrolyte levels are lower than the amount the body needs to maintain healthy function, or, conversely, when the concentration of a given electrolyte increases beyond what the human body can regulate effectively (Felman, 2017). The most common electrolyte imbalances involve sodium or potassium (Felman, 2017). The actual symptoms of electrolyte imbalance conditions are determined by two main factors. The first factor is the actual electrolyte that is out of balance. This is because, as mentioned previously, various electrolytes serve to fulfil different functions in the human body. The second factor that determines the symptoms of electrolyte imbalance conditions is whether the levels of that particular electrolyte are too high or too low.

For example, excessive magnesium, sodium, potassium, or calcium can result in irregular heartbeat (arrhythmias), confusion, changes in mean arterial pressure, bone disorders, twitching, nervous system disorders, muscle spasms, and tiredness (Felman, 2017). Calcium excess in particular is especially common in patients with pre-existing health conditions such as lung carcinoma, breast carcinoma, or multiple myeloma. Additional symptoms of this type of electrolyte imbalance include fatigue, lethargy, frequent urination, coma, loss of appetite, and constipation (Felman, 2017). However, the difficult part about calcium excess is that because it is a common condition in cancer patients, and cancer treatments can result in much of the same symptoms as calcium excess, it can be very tricky for healthcare practitioners to recognize that a patient is experiencing high calcium levels (Felman, 2017).

When electrolyte levels are too low, healthcare practitioners will often prescribe the required electrolyte through supplements; however, this, of course, depends on how severe the imbalance is (Felman, 2017). According to the World Health Organization (WHO), oral rehydration therapy, a therapy used for people who exhibit symptoms of electrolyte shortage alongside other symptoms, such as dehydration (which will normally be the after-effect of a bout of severe diarrhea), is administered. The approved solution by the WHO includes 2.6 grams of potassium, 1.5 grams of potassium chloride, and 2.9 grams of sodium citrate. All of these electrolytes are dissolved in a singular liter of water, and administered to the patient orally. On the other hand, electrolyte replacement therapy is used when a patient exhibits signs of a much more serious electrolyte shortage. This type of therapy can be administered orally as well, but it can also be given to patients via an intravenous (IV) drip. For example, extreme sodium deficits are treated via electrolyte replacement therapy by administering an infusion of saltwater solution or compound sodium lactate through an IV in a hospital setting.

Electrolytes are crucial for athletes' sporting performance because when muscles move, it is the electrical impulses travelling through them that facilitate all of their function (Elete Electrolyte, n.d.). These electrical impulses do not simply appear out of nowhere! Rather, it is all of the electrolytes inside of the human body that generate the electrical impulse that is conducted through athletes' muscles. As a result, when electrolyte levels are too low in an athlete's body, due to the athlete sweating them out, muscles can start cramping or spasming, because the electrical impulse is not being carried through their muscles correctly. Cramping is essentially the body's way of telling an athlete that electrolyte levels are down (Born, n.d.)

Athletes in particular are extremely prone to electrolyte deficiencies, simply because they need more electrolytes. In fact, competitive athletes are considered to be "predestined" to evolve magnesium deficiency as a result of increased sweating and urination from their training. This can be detrimental to the athletes' health because magnesium, as stated previously, is a necessity for the muscles and other organs. It is for this reason why magnesium supplements are often used, and in fact, can be encouraged by some as crucial measures during long periods of physical stress (Born, n.d.)

The conversation pertaining to the science behind electrolytes cannot be discussed without mentioning homeostasis (Elete Electrolyte, 2017).

Homeostasis is the process that involves the human body maintaining fluid levels of electrolytes within highly specific and narrow ranges. For example, when the body has an electrolyte deficiency, the electrolyte will be absorbed at higher frequencies via the digestive system as an attempt to retain it. Further, less of the electrolyte will be excreted via the sweat or the urine. It is for this reason why athletes with sodium or magnesium deficiencies might notice that their sweat has lost its characteristic salty flavour; the salt is being retained by the body to fight against the electrolyte deficiency (Elete Electrolyte, 2017).

However, it is of utmost importance to note that sodium, as mentioned prior, is the most common electrolyte found in the modern diet (Elete Electrolyte, 2017). Thus, most peoples' sweat is actually quite high in sodium content, but conversely very low in potassium and magnesium. This is why clinical studies that use sweat as a proxy for understanding which electrolytes need to be replaced versus which ones exist in excess are not the most reliable.

The human body can replace, at most, roughly a third of the electrolytes, calories, and fluids that it will lose during vigorous exercise (Born, n.d.). However, that does not mean the athlete should take it upon him or herself to replace all of the lost fluids and electrolytes at once. In fact, electrolytes lost are not simply replaced by the electrolytes one consumes in the moment. Rather, if an individual attempts to replace all of the fluids - and, by consequence, the electrolytes - their body has excreted all at once, it is possible to acquire dilutional hyponatremia, a condition that involves the blood sodium levels becoming overly diluted. This type of oversupplmentation of fluid can result in water intoxication (Born, n.d.). Similarly, replacing all of the electrolytes an individual presumes they have lost immediately following vigorous exercise can result in hormonal triggers and further imbalances, which can result in edema, muscle spasms, cramps, as well as distress to the gastrointestinal system (Born, n.d.). As a result, it is encouraged for athletes to fuel and refuel themselves based on how much their bodies can accept and absorb, rather than overloading the human body with electrolytes and fluids immediately following exercise.

In summation, the science behind electrolytes is quite simplistic, yet it is a significant driving force in the human body that enables people, especially athletes, to carry out their bodily functions. Electrolytes themselves are simply ions composed of positively charged and negatively charged mineral complexes that can dissolve and separate in bodily fluids. The separation and dissolution of these charges facilitates the conduction of electrical signals throughout the body, which consequently aid in the intricacies of muscle function, nervous control, fluid balance, pH balance, and various other processes. Athletes are prone to electrolyte loss because exercise causes a change in one's fluid and electrolyte balance. The main reason for this is sweating, due to increased arterial pressure, allowing for more electrolytes to be lost via the sweat glands. However, although it may seem like the best solution to this excessive electrolyte loss is supplementing with electrolytes after exercise, this might not necessarily be a good idea, as it can lead to more complicated health conditions.

CHAPTER 7.

Do Electrolyte Drinks 'Work'? Are Electrolyte Drinks Reliable? Are Electrolyte Drinks Better Than Water?

Alexa Gee

Introduction

Athletes looking to boost their physical activity commonly turn to energy drinks. These drinks typically contain the electrolytes, sodium and potassium, and can have sugars added to the mixture as well (Ostojic & Mazic, 2002). When exercising, muscle glycogen is depleted at around two hours of intense activity (Davis et al., 1988). Muscle fatigue happens with the depletion of this glycogen and the reduction in blood glucose concentration (Ostojic & Mazic, 2002). Through sweating, sodium and potassium are lost, depleting the stores of these electrolytes within the body (Powers et al. 1990). The consumption of carbohydrate-electrolyte drinks replenishes glucose for oxidation in the muscles, and the previously mentioned electrolytes, delaying the onset of fatigue.

Do electrolyte drinks work?

These drinks have been tested with various types of exercise styles and intensities. The volume of electrolyte beverage consumed should be greater than the volume of sweat loss to replenish loss electrolyte stores (Maughan et al., 1997). A higher concentration of sodium is lost more than magnesium or potassium ions in sweat (Dragos-Florin, 2017). The lack of electrolytes can induce muscle cramps so electrolyte drinks have been advocated by health and sports professionals to restore balance (Miller et al. 2009). Magnesium and calcium ions in electrolyte drinks will also help with muscle contractions (Raizel et al., 2019, p. 5-10). However, Miller et al. (2009) find that the consumption of electrolyte drinks do not offset the effects of muscle cramps nor do they significantly change plasma electrolyte concentrations compared to the water control after one hour of exercise. As evidenced by this study, electrolytes do not always work in this capacity.

Electrolyte drinks can be consumed pre-workout or post-workout.

Consumption of the drinks pre-workout delays fatigue, allowing improved physical performance. Before exercising, drinking sodium electrolyte drinks can help retain water prior to sweating (Shirrefs et al., 2007). Doing so will offset significant losses of sodium (3-4 grams) in sweat when exercising. This ensures that the athlete does not start exercising with a negative balance of electrolytes, which would further decrease performance (Watson et al., 2008). Ingesting the electrolyte drink, Gatorade, 20 minutes before exercise increases short-term performance and limits lactate accumulation in the bloodstream (Singh et al., 2011). The sodium content in Gatorade maintains blood volume by reabsorbing more water in the intestine. This lowers heart rate and causes greater blood flow to the muscles and skin, improving cardiovascular response and delaying fatigue.

While working out, it is suggested that athletes should ingest between 30 to 60 grams/hour of carbohydrate drink to maintain electrolyte balance (Watson et al., 2008). In different research studies, electrolyte drinks improve cycling times, lower heart rate when running, and allow soccer players to dribble the ball faster (Davis et al., 1988; Roberts, 2017; Ostojic & Mazic, 2002). Similarly, McRae and Galloway establish that ingestion of carbohydrate-electrolyte drinks improve serving and return-of-serves while playing tennis for 2 hours, implying that electrolyte drinks can improve visual reaction time (McRae & Galloway, 2012). Contrary to the other studies, the electrolyte drink did not improve coordination or a power test when kicking a soccer ball (Ostojic & Mazic, 2002).

After a workout session, electrolyte drinks rehydrate the body and replenish electrolytes lost from sweating (Dawes et al., 2014). Electrolyte drinks can also be used for those experiencing illness, excessive diarrhea like in the case of cholera, or those that do not consume enough potassium or sodium in their diet (Busch, 2018). A severely low sodium level in the bloodstream is termed hyponatremia (Raizel et al. 2019, p. 5). If sodium levels are not returned to normal levels, a coma or even death can result if left untreated (Bhargava, 2021). Electrolyte drinks are a way to rectify this. Ingestion of at least 18 mmol of sodium prevents dehydration (Singh et al., 2011). Effective rehydration after intense long-term (more than one hour) exercise requires at least 50 mmol/L of sodium and smaller amounts of potassium and carbohydrates (Maughan et al., 1997). Sodium improves the rate of intestinal uptake water. The sodium in carbohydrate-electrolyte mixtures also helps with the delivery of glucose to the muscle cells (Powers et al., 1990). A study by Powers et al. (1990) found that those who drank a 12.5 g/L solution of carbohydrate-electrolyte drinks had a significantly higher blood glucose level after 20 minutes post-exercise compared to participants who drank the control, which contained no electrolytes. These different cases demonstrate that electrolyte drinks are capable of enhancing performance, provided the right amounts are ingested.

Are electrolyte drinks reliable?

The reliability of electrolyte drinks depends on a variety of factors. Electrolyte concentration varies widely between brands (Coombes & Hamilton, 2000). Gatorade contains 110 mg/250 mL of sodium, whereas Powerade contains 70 mg/250 mL and Hydrafuel only contains 25 mg/250 mL (Leiper, 1998). Since the composition of electrolytes varies between different drinks, they will have different effects on the body (Nielsen et al., 1986). For example, Glucose-electrolyte solutions restored plasma volume significantly faster than water. Sports drinks contain 6-8% carbohydrates, sodium and potassium levels are much lower than this (Coombes & Hamilton, 2000). A solution of 7.6 g/L of carbohydrate-electrolyte solution did not cause significant replenishment of plasma sodium, potassium, calcium, or chloride. Ingestion of water showed the same amount of plasma concentration increases (Jeukendrup et al., 1997). These beverages did not contain enough electrolytes to make a marked difference in plasma concentration. For 1-2 hour sessions of endurance activities, electrolyte drinks containing 30 grams of carbohydrates or more were found to be beneficial. Sugar and sodium are essential components of electrolyte drinks like sports drinks; the body absorbs water better in the presence of glucose and sodium ions help with the retention of water in the intestinal wall (Leiper, 1998). The trend in sugar-free energy drinks is counterproductive as sugar is needed for the drink to work effectively because of the former reason. High sodium drinks favour the filling of the extracellular compartments, increasing plasma volume, while high potassium drinks favour intracellular rehydration. Shirreffs et al. (2013) found that sodium is the most important electrolyte for water retention; without it, large amounts of water would be excreted as urine. There is less evidence to suggest that potassium and magnesium givelarge benefits when added to electrolyte drinks. Although, between high sodium and high potassium drinks, there is no effect on mean heart rate. High potassium drinks do not result in higher fluid retention than sodium and do not pose any additional rehydration benefits over sodium (Perez-Idarraga & Aragon-

Vargas, 2014). High potassium solutions do not facilitate more glycogen build-up, having low glycogen means that there is a reduced capacity for physical activity, meaning potassium does not likely inhibit fatigue. An additive effect of potassium and sodium on rehydration has not been found (Maughan et al., 1994). In actuality, there is a slower rate of plasma volume restoration in an electrolyte drink containing both sodium and potassium compared to a drink with just sodium (Nielsen et al., 1986).

At moderate intensities and activity of more than two hours duration, carbohydrate-electrolyte drinks improve cycling endurance performance and cycling time (Jeukendrup et al., 1997). For upper body muscular endurance performance, electrolyte drinks are beneficial (Dawes et al., 2014). When doing push-ups or sit-ups, electrolyte drinks enhance performance by 12 and 13% respectively. Yet, other studies have shown that electrolyte beverages fail to delay the onset of fatigue after high-intensity cycling (Nassis et al., 1998).

The benefit of electrolyte drinks depends on the type of exercise done (Hornsby, 2011). Electrolyte drinks are unnecessary for those who do low-intensity, short workouts (Meixner, 2018). For short-term exercises, like 10 minutes of cycling, 6% and 9% carbohydrate-electrolyte drinks did not enhance performance, lower heart, or decrease exertion. It was not found to impede this type of exercise either. Therefore, electrolyte drinks have no effect on short-term exercise since electrolytes are not severely depleted at this level of physical activity (Meixner, 2018). Additionally, weight-training exercises usually do not deplete carbohydrate stores as much as endurance so consuming energy drinks here also has no added benefit (Meixner, 2018). If sedentary individuals or non-athletes consume electrolyte drinks, there is no benefit either and this beverage is usually just an added source of sugar.

Energy drinks do not work indiscriminately. Their functions depend on a variety of factors internal and external to the body. Electrolyte losses decrease as the training level of the athlete increases; highly trained athletes' bodies are able to minimize the amounts of sodium and potassium excreted in their sweat. Thus, electrolyte concentration variation between the brands was found not to matter as much for elite athletes as opposed to more casual ones (Leiper, 1998). Aside from the level of physical activity performed, the bodyweight of individuals also needs to be taken into account (Chatterjee & Abraham, 2019, p. 533). Moreover, most studies have only looked at samll focus groups of young biological males, so there could be differences in the demand for electrolytes in different age groups or genders other than cisgender males.

Is it better than water?

Water or electrolyte drinks can be used to rehydrate the body after low-intensity, short-term work-outs. Water was found to be the best fluid replacement while playing tennis in the heat for moderate or low-intensity activities that were less than one hour (Dragos-Florin, 2017). In contrast to the previous study, compared to water, electrolyte drinks significantly increased biking distance and step number in a short-intensity period (25 minutes) of cycling (DiSilvestro et al. 2011).

In another study with military participants, carbohydrate-electrolyte drinks did not improve target shooting with rifles compared to water after an exercise involving a 700-metre run (Tharion et al. 1995). This suggests that electrolyte drinks provide no additional benefit for fine motor tasks. Mental performance, cognition, is similarly unaffected by carbohydrate-electrolyte drinks, insignificantly different from results tested with water (Meckes & Brown, 2017).

For high-intensity exercise, like running on the treadmill for more than one hour, electrolyte drinks are recommended instead of consuming water (Kalman et al. 2012). Marathon runners drinking a 5.5% carbohydrate-electrolyte solution had better performance times than runners who drank water (Tsintzas et al., 1995). Tap water (as opposed to pure water) contains trace amounts of minerals and nutrients dissolved in it (Tinsley, 2018). One litre of tap water contains 2-3% of the reference daily intake (RDI) for sodium, calcium, and magnesium, but has a very low concentration of potassium. This is not enough to placate the demand after an intense workout session. In fact, drinking pure water can be dangerous. As mentioned previously, hyponatremia can happen when the body excretes too much sodium from sweating (Bhargava, 2021). Marathon runners are likely to experience this if they only drink water throughout the course of their run, diluting their body fluids (Dragos-Florin, 2017). Drinking beverages with sodium triggers the thirst mechanism causing the person to consume more fluids and ensures more sodium enters the bloodstream (Raizel et al. 2019, p. 5-10). Unfortunately, Rosner and Kirven (2007) found that sports drinks had insufficient levels of sodium to prevent hyponatremia. Most sports drinks have a sodium content of 230-460 mg/L (10-20 mmol/L) (Rosner & Kirven, 2007). The American College of Sports Medicine suggests a sodium content that is around two times higher than this number to replenish what was lost by sweat (Rosner & Kirven, 2007).

If exercising in a humid and hot place, where sweating can be quite excessive, electrolyte drinks are also recommended over water (Cleveland Clinic, 2021). The body's mass is lost through sweating, aerobic performance begins to decline (Shirreffs et al., 2007). It was found that basketball players attempted fewer shots and scored less frequently once 2-4% of the body mass is lost. 20-30mM/L of sodium and 2-5mmol/L of potassium are recommended for rehydration (Shephard, 2019, p. 148). On average, electrolyte drinks contain 18% RDI for sodium, 3% for potassium but typically low concentrations of magnesium or calcium. There are different types of electrolyte drinks, designed for different levels of physical endurance, where a mixture for ultra-endurance activities (exceeding six hours) would have greater concentrations of potassium or sodium (Cleveland Clinic, 2021; Wortley & Islas, 2011).

Electrolyte drinks can also be applied as buffers in the bloodstream, whereas tap water does not have this property (Powers et al., 1990). A fluid-replacement beverage that contains sodium citrate/citric acid minimizes alteration in H+ ions in the blood during intense exercise, stabilizing blood pH. Blood lactate levels are lowered (Khanna & Manna, 2005). Buffering the buildup of lactic acid is an ergogenic benefit since it enhances recovery, although this ability has not been found in all electrolyte drink brands (Powers et al., 1990; Roberts, 2017). The sports drink brands Gatorade, Powerade had an ergogenic effect but Aquarius did not contain enough sodium to have this benefit (Del Coso et al., 2008).

An advantage water has over carbohydrate-electrolyte drinks is that water is the healthier alternative (Sampaio de Melo et al., 2016). It was found that electrolyte drinks cause erosion of the tooth enamel surface, due to low pH (<5.5) and high sugar content. For this reason, tap water is recommended instead when water and electrolyte drinks would provide an equal benefit to the consumer. Coconut water has been looked at as an alternative to manufactured electrolyte drinks, which usually contain high amounts of fructose and artificial flavouring (Kalman et al., 2012). Coconut water does have a low pH but contains a high concentration of calcium, which inhibits the tooth erosion process (Sampaio de Melo et al., 2016). Furthermore, coconut water shows comparable effects of fluid retention to sodium and potassium-filled electrolyte drinks and tap water (Kalman et al., 2012).
Interestingly, milk is another viable alternative to sports drinks (Desbrow et al. 2014). Cow milk naturally contains high amounts

of sodium, potassium, and milk proteins; these ingredients aid in rehydration after an intense workout (Desbrow et al., 2014; Watson et al., 2008). Milk proteins can also contribute to the replenishment of muscle glycogen stores (Watson et al., 2008). Cow, skimmed, and soy milk were all found to improve fluid retention more so than carbohydrate-electrolyte drinks after bouts of moderate intensity workout periods (Desbrow et al., 2014; Watson et al. 2008). More data on the applicability of milk for high intensity exercise is needed.

Summary

Electrolyte drinks have grown in popularity as they have shown to be successful at improving various kinds of physical activities. Water has been studied as an alternative to consuming these electrolyte beverages. Electrolyte drinks show the greatest benefit at rehydrating the body and delaying fatigue for endurance activities that last longer than one hour.

CHAPTER 8

How and Where Electrolytes are Produced and Excreted in the Body

Alexia Di Martino

Electrolytes play a major role in maintaining various systems throughout the human body. Cells are riddled with various transmembrane voltage-gated channels where electrolytic ions can move through extracellular and intracellular spaces by way of active or passive transport. To maintain these precisely regulated systems, electrolytes must be able to enter, move through, and exit the complex system of the body.

Production of Electrolytes

Electrolytes are incapable of being organically synthesized in the human body. Thus, they enter the body through dietary consumption. After sufficient processing and absorption into the correct systems and organs, the electrolytes will then be used to carry out electrical and chemically charged functions.

Ingestion of foods or fluids containing various electrolytes is a way for ions to enter the body. The subsequent digestion and absorption allows the ions to enter the bloodstream. The bulk of electrolytes consumed typically comes from solid, gastrointestinal-digested food, but can also be from specially formulated electrolyte drinks. Although consumption occurs in the mouth, the long digestive process does not see the consumed electrolytes get incorporated into the body until nearly the very end in the intestines. "All segments of intestine from duodenum to distal colon have mechanisms for both absorbing and secreting water and electrolytes" (Field, 2003). The gastrointestinal tract will filter through about 8 litres of electrolytically concentrated fluid from the digestive process every day (Allison et al., 2004). Concentrations of this fluid within the intestines typically contain around 800 mmol of sodium, 700 mmol of chloride, and 100 mmol of potassium, and this electrolytic fluid is processed primarily by the small intestine (Kiela &

Ghishan, 2016). The small intestine is subdivided into three smaller parts – the duodenum, jejunum, and ileum (Collins et al., 2020). The jejunum plays a major role in electrolyte digestion and absorption. This is supported by the villi and microvilli, which are hair-like projections protruding from the absorptive intestinal epithelial cell surface (Kiela & Ghishan, 2016). Brush border digestive enzymes are found within these microvilli on the apical surface of the intestinal cells (Hooton et al., 2015). These enzymes work to break down larger molecules of food in the intestine into smaller components. Enzymatic digestion further frees up the ions to move around in the intestinal lumen spaces. This is the stage where electrolytes can finally enter the bodily systems for functional purposes. Each electrolyte has specific methods of transport or certain mechanisms that they use to pass through the intestinal lining; Sodium has a wide variety. There are sodium-dependent nutrient transporters that allow sodium to perform nutrient-coupled sodium absorption by co-transporting molecules such as glucose. There are also exchangers embedded in the intestinal lining such as sodium-hydrogen protein exchangers, which trade a hydrogen ion inside of the serosa on the basolateral membrane for the sodium in the lumen. A final simple example is sodium channels in the epithelium of the intestines that permit passive sodium absorption (Kiela & Ghishan, 2016). Similar mechanisms exist for the other electrolytes. Chloride is absorbed from the lumen of the intestines through passive transport or simple diffusion through the intestinal epithelium down its concentration gradient and out of the lumen. It can also use a chloride-bicarbonate exchanger, which brings a bicarbonate ion into the lumen in exchange for a chloride ion (Kiela & Ghishan, 2016).

Planned ingestion of specific levels of electrolytes can be necessary to maintain bodily ion balance. Later in the chapter, quantities of ion excretion will be discussed. The recommended daily intake of each type of electrolyte is based on their daily excretion levels to ensure that what is being lost each day is replenished to support all electrically charged bodily processes. The minimum dietary requirements of ions needed are far exceeded by our typical diet (Gerritsen et al., 2015), which has likely contributed to the increased prevalence of disorders such as high blood pressure (Betts et al., 2013). An adult not partaking in strenuous exercise leading to sweating would need only 100 mg of sodium per day to replenish that of which is lost in daily excretion (National Research Council, 1989). Childhood development and pregnancy can see massive changes in sodium expectations. Pregnancy can see a person gain about 11 kilograms of body weight, which increases the daily sodium requirement by 69 mg. Lactation is also a very sodium-requiring process, adding about another 135 mg to a pregnant person's daily diet to a total of 304 mg compared to a typical adult's 100 mg requirement (National Research Council, 1989).

Storage of Electrolytes
After electrolytes have been absorbed, but before they are excreted, they must exist elsewhere in the body for later use. This leads to the body's various storage systems of electrolytes. Allison et al. (2004) discussed the retention of electrolytic fluids, stating that the retention of bodily fluid in cellular compartments affects the internal ion balance between these compartments.

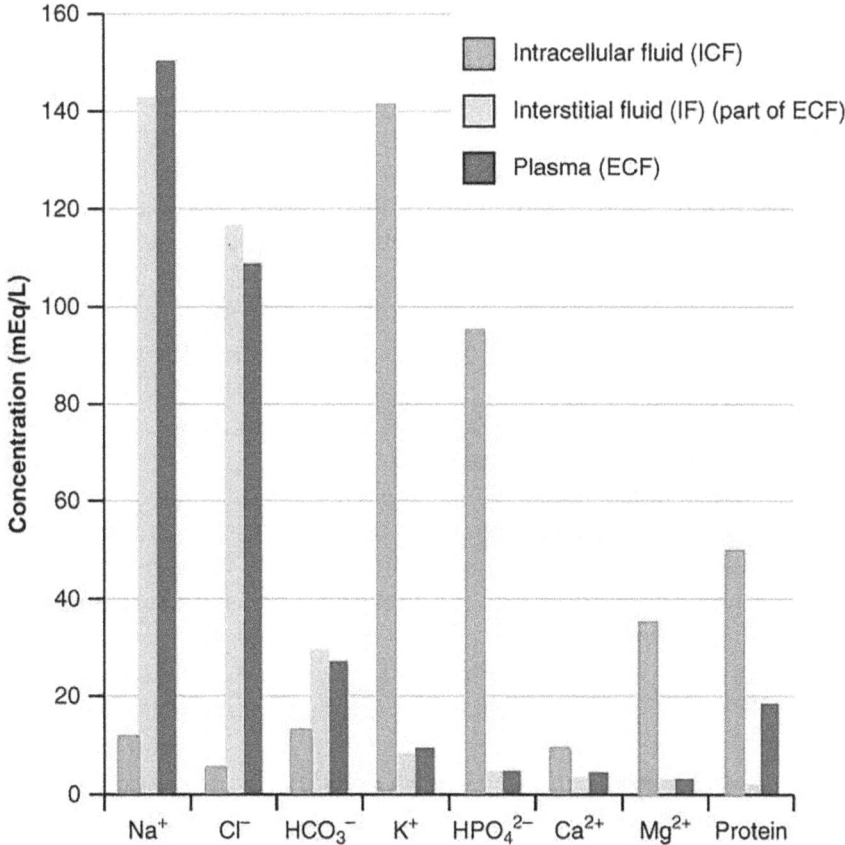

Source extracted from OpenStax, 2018

The intracellular fluid and extracellular fluid are major stores of many substances in the body in addition to electrolytes. These two types of fluid account for two-thirds of the total body water. The intracellular fluid refers to all cytoplasmic fluid within bodily cells that surrounds organelles encompassed by each cellular membrane and it comprises 60% of bodily water. Extracellular fluid covers a much wider range of fluids, accounting for about one-third of bodily water (OpenStax, 2018). It is the fluid that exists around all cells in the body and refers to two smaller divisions of fluid – blood plasma and interstitial fluid. Figure 1 describes the proportions of electrolytes that make up both types of fluids. Blood plasma is highly concentrated, containing a great amount of sodium, chloride, and bicarbonate. Intracellular fluid contains a great amount of potassium electrolytes, while extracellular fluid only contains trace amounts of it.

The presence of electrolytes in the blood provides a wide array of possibilities for it in other spaces in the body. Blood flow is crucial for the survival of all tissues. Through the movement of blood and fluids throughout the body, comes transport of these dissolved electrolytes as well. In relation to Figure 1 and the concentrations discussed, movement of blood through capillaries will carry the potassium of the intracellular red blood cell fluid, as well as the sodium, chloride, and bicarbonate electrolytes of the blood plasma. Under the influence of osmotic and hydrostatic pressure, fluid can move between capillaries and vascular space bringing electrolytes with them all over the body. Many cellular membranes also have selective permeability, allowing sodium and chloride ions to freely diffuse between the plasma and interstitial fluid (Darwish & Lui, 2021).

The shifting of electrolyte storage to other bodily compartments through electrolyte transport in capillaries is a contributing factor of many human illnesses such as cholera. A patient diagnosed with cholera may experience fluid and electrolyte loss from the mucosal layer lining the inside of their intestines into the intestinal lumen (Banwell et al., 1970). Substances in the lumen comprise what will eventually be excreted as feces. High lumen fluid concentrations will result in watery stool, or diarrhea, which is characteristic of a cholera infection. The study by Banwell et al. (1970) also found that this fluid had an osmolarity that was very close to that of the blood plasma extracellular fluid. This means that cholera patients experience a severe electrolyte imbalance due to the rapid excretion of sodium, potassium, and chloride ions. There was also a high concentration of bicarbonate ions found in the lumen of cholera patients, which can lead to the development of acidosis (Wang et al., 1986).

Excretion of Electrolytes

Electrolytes are naturally excreted through typical bodily function. The main route by which ions can exit the body is through filtration by the kidneys and into the urine. However, sweat and feces also play minor roles in smaller proportions of ion excretion.

The kidneys are highly compartmentalized organs, and the production of urine is a highly regulated process. Two hormones in particular, aldosterone and angiotensin II, are crucial and must be released to begin the urine formation process. Aldosterone is released by the adrenal gland located on top of the kidneys in response to a change in electrolytic concentrations of the blood. This can be conditions such as an increase in blood potassium levels or a sharp decrease in blood sodium levels. Upon recognition of these conditions, angiotensin II stimulates the secretion of aldosterone. Aldosterone works to influence the kidney to reabsorb sodium and water into the blood, reducing the excretion of sodium electrolytes in urine (Scott et al., 2021). The increase in sodium concentration in the blood creates an osmolarity gradient, causing water to follow sodium into the capillaries to establish isotonicity. However, aldosterone coordinates this process by utilizing sodium-potassium pumps in the functional units of the kidneys - the nephrons (OpenStax, 2013). This means that as sodium is pumped back into the tubules for reabsorption into the blood and capillaries, potassium electrolytes are being pumped out at an equal rate into the nephron's filtrate to be excreted by the body in urine.

Urinary excretion of electrolytes allows the precisely regulated exit of sodium, potassium, and chloride ions. On average, human adults excrete 1558 ± 689 mL during a 24-hour period (Wang et al., 2013). Within this urine, the total mass of each kind of excreted electrolyte varies. During the 24-hour collection period of a study by Wang et al., about 3.3 grams of sodium, 1.98 grams of potassium, and 4.85 grams of chloride were excreted on average per person. As found by the work of Wang et al. (2013), the time of day of urination can affect the concentration of excreted electrolytes. Overnight samples of urine collected were found to be higher in volume than those collected at other times in the day. The urine samples collected at night also contained a lower concentration of sodium, potassium, and chloride ions than those during the day. No bicarbonate ions exit the body through the urinary tract, as the kidneys do not filter it out. Instead, they preserve it for later use in buffering systems to maintain the pH of bodily fluids such as blood (Betts et al., 2013).

Excretion of electrolytes can also occur through the intestinal tract by production of feces or vomit. In a day, the human gut and intestinal tract can filter through approximately 8 litres of fluid per day through the bowels (Allison et al., 2004). After filtration, 6.5 litres will have been absorbed back into the body (Field, 2003). Excretion of ions in the intestine occurs in multiple causative steps. To begin, cyclic AMP (cAMP) stimulates cystic fibrosis transmembrane conductance regulator (CFTR), an anion channel on the apical surface (facing the intestinal lumen) of the intestinal epithelial cells which allows the negative chloride ions as well as bicarbonate to enter the lumen. This loss of negative ions inside the epithelial cells is counteracted by an influx of potassium from the serosa through potassium channels in the basolateral membrane. This allows the internal cell polarization to remain the same. The movement of the chloride creates a potential difference between the serosa and the lumen. Sodium cations are already present in the serosa due to the capillary blood supply running adjacent to it, creating a positive membrane potential on the basolateral side. Simultaneously, chloride anions are being transported into the lumen, creating a negative membrane potential on the apical side. The intestinal epithelial cells contain intercellular tight junctions between them, through which sodium is able to move into the lumen along the electrical gradient and equalize the difference in membrane potential (Field, 2003).

Skin is the largest organ in the body. Not only does it provide protection from external factors such as pathogens, injury, and UV light, it also plays a role in the excretion of electrolytes. Human skin is composed of three layers that are anatomically distinct and perform unique functions. The epidermis is the uppermost layer which is followed by the dermis and hypodermis (Yousef et al., 2020). The dermis layer contains thermoreceptors that monitor thermal conditions and send signals back to the brain. The core temperature of the body can increase considerably in hot environments or during intense exercise due to muscular contraction. To maintain homeostasis, this excess heat must be dissipated. Sweating is a process that allows for thermoregulation through water loss. By secreting water onto the surface of the skin, the excess body heat can transfer to the secreted water, causing evaporation. This evaporation creates a cooling effect that results in heat loss (Baker, 2017). There are two types of sweat glands – eccrine sweat glands and apocrine sweat glands. Sweat glands are embedded in the dermal layer of the skin and extend upward to the epidermis, opening up to either a hair follicle – in the case of apocrine sweat glands - or to a sweat pore on the surface – in the case of eccrine sweat glands (Baker, 2018). During perspiration, these sweat glands will release water from inside the body to these openings on the surface of the skin (Hodge et al, 2020). The primary sweat secreted contains a similar electrolyte concentration to blood plasma, of around 140 mmol/L of sodium, 100mmol/L of chloride, and 5 mmol/L of potassium (Baker, 2017). For reference, Figure 1 shows typical blood plasma electrolyte concentration. Electrolyte reabsorption occurs during the journey of the primary sweat from the gland deep in the dermis up to the epidermis and outside of the body. Sodium reabsorption occurs through both passive and active transport, with epithelial sodium channels and sodium-potassium ATPase transporters, respectively. Chloride ions are reabsorbed into the body through CFTR proteins in a passive mechanism (Baker, 2017). These channels are embedded in the skin membranes. From these electrolyte reabsorption processes, the sweat that actually reaches the surface of the skin is hypotonic and has a much lower osmolarity.

The well-known autosomal inherited disorder, cystic fibrosis (CF), is typically discussed in relation to the mucosal respiratory effects that it causes. CF is lesser known for its complications of sweating in those with the disorder. CFTR is a protein present in both CF patients and unaffected individuals that, among other functions, aids in the ion regulation of epithelial chloride ion channels in the sweat gland. In unaffected individuals, the sweat gland is able to reabsorb chloride through its ion channels, bringing sodium along with it in the coupled messenger pathway for reabsorption as well. The final sweat that evaporates from the skin is highly dilute, minimizing electrolyte loss. Due to mutations leading to dysfunction in the CFTR of CF patients, the sweat gland chloride ion channels are incapable of reabsorbing chloride. This leads to the lack of coupled reabsorption of sodium and chloride, causing secretion of highly ionically concentrated sweat and a hypotonic intercellular fluid (Hodge et al., 2020). As a result, CF patients can unfortunately experience extreme electrolyte imbalance when they sweat, which should be a beneficial and regulatory process.
In cases of pregnancy, electrolyte excretion increases in order to support a variety of reproductive functions and fetal development. The electrolyte concentration in the amniotic fluid is particularly important and must be regulated to avoid developmental complications,

or even the presence of a related disease such as hypertension later in the developing baby's adult life (Sakuyama et al., 2016). Lactation is a reproductive process by which the mother can provide nutrient-rich milk from the breasts to their child in the early months of life. The excretion of this milk through lactation is linked to the transport of electrolytes. Human milk contains 390mg chloride per litre, 500 mg of potassium per litre, and 180mg sodium per litre (National Research Council, 1989). As seen earlier in this chapter, children have different and increased dietary demands for electrolytes. These concentrations in human milk likely evolutionarily correspond to the electrolyte requirements of an infant entirely dependent on breastfeeding to obtain nutrients.

Conclusion

The tendency to homeostasis in all human systems drives the absorption, storage, and excretion of electrolytes in the body. From intracellular fluids to whole organ systems, electrolytes are highly present and concentrated all throughout the human body. While electrolytes can only be sequestered through dietary means, their excretion varies widely. Following basic mechanical and chemical digestion, electrolytes enter the body from the lumen of the small intestine using various mechanisms of active and passive transport. Illnesses and disorders such as cholera and cystic fibrosis can affect ion management as well. In the cases of most healthy individuals, as long as dietary ingestion of sufficient sodium, potassium, and chloride is occurring, the body can take care of its absorption, transportation, and excretion processes with high accuracy.

CHAPTER 9

False Electrolyte Advertising

Amna Zia

Background: Electrolyte/Sports Drinks Market

Electrolyte drinks, commonly known as sports drinks, are beverages designed to replenish the body's carbohydrate stores, fluids, and electrolytes (sodium, potassium, magnesium, calcium) that are lost during vigorous physical activity (Pound & Blair, 2017). The carbohydrate content of sports drinks typically comes from sugar sources such as glucose, sucrose, fructose or maltodextrin, although some sports drinks are designed low- or zero-carb for consumers who want the hydration and electrolytes without the added calories (Pound & Blair, 2017; Tinsley, 2018).

By 2026, the global sports drink market is expected to reach $32.61 billion, with the North American market standing at a grand $7.90 billion (USD) as of 2018 (Fortune Business Insights, 2021). The market for sports drinks in North America has become immensely concentrated due to the growing number of health-conscious consumers in the region (Fortune Business Insights, 2021). The key market drivers in the North American sports drink industry are PepsiCo and The Coca-Cola Company — the former manufacturing the well-known brand, Gatorade and the latter manufacturing its rival, Powerade (Conway, 2021). Recent market trends indicate that sports drinks also compete with non-sport beverages such as bottled water, enhanced water, coconut water and even chocolate milk with respect to being the best solution for post-workout dehydration (Conway, 2021).

There have been multiple lawsuits and case proceedings involving sports drinks and false advertising over the past decade (e.g., Ackerman v. The Coca-Cola Company) (Brison et al., 2020). This is because in order to stand out amongst the increasingly competitive marketplace, sports drinks companies make strong subjective claims about the product's health or performance benefits and have prominent athletes endorse their products to legitimize the product's perceived efficacy (Brison et al., 2020). Moreover, sport drinks companies also heavily rely on "selling the science", i.e., the pairing of scientific credibility with aggressive advertising (Cohen, 2012). Since the 1990s, many sports drink manufacturers have infiltrated the area of clinical science and have since developed long-standing financial affiliations with medical groups, academic researchers and sports associations (Cohen, 2012). These ties have allowed the sports drink industry to condition the public to believe that sports drinks are among the most extensively researched food products and that fluid consumption is equally vital for athletic performance as exercise training — beliefs that are often accepted with little consumer skepticism or scrutiny (Cohen, 2012).

It is imperative that consumers are aware of sports drinks advertising that is misleading and/or omits important information as exposure to such marketing may influence them to develop uninformed beliefs about a product, and greatly overestimate its nutritional, health or performance benefits. To aid this understanding, this chapter will discuss cases involving deceptive advertising of products containing electrolytes.

BodyArmor SuperDrink and BodyArmor Lyte

The National Advertising Division (NAD) — a national program that monitors the authenticity of advertising in all media across the U.S. — found that BodyArmor Nutrition made unsubstantiated claims in the advertising of their BodyArmor SuperDrink and Body Armor Lyte sports drinks (BBB National Programs, 2020). After reviewing the company's social media videos, banner ads, in-store displays and a press kit, the NAD requested that BodyArmor Nutrition discontinue or modify the claims that the BodyArmor SuperDrink is "The only sports drink. No artificial sweeteners, flavors or dyes. Potassium packed electrolytes"; and that the BodyArmor Lyte is "The only sports drink. Low calorie. No sugar added. No artificial sweeteners, flavors or dyes" (BBB National Programs, 2020). The NAD concluded that these claims were hyperbolic as they imply that the BodyArmor SuperDrink and BodyArmor Lyte are the only sports drinks of their kind to exist, when in fact other sports drinks such as Gatorade also possess the same aforementioned qualities, e.g., being packed with electrolytes, having no artificial sweeteners, etc. (BBB National Programs, 2020). The NAD expressed concern that consumers may take away a false message with regards to the product's "comparative superiority to other sports drinks" (Edelstein, 2018).

Coconut Water

In 2012, a Canada-wide class action lawsuit was launched against the makers of Vita Coco, one of the top-selling coconut water brands, for deceptive advertising (Consumer Law Group, n.d.; Theeboom, 2014). According to the lawsuit, Vita Coco claims its products to be "super hydrating", "nutrient-packed", "mega-electrolyte", and "life-enhancing", though independent studies have found their products to contain far less electrolytes and nutrients than stated on the product label (Rothstein, 2012).Tests showed that Vita Coco's coconut water had 40% less sodium, 35% less magnesium, and 16% less potassium than advertised on the product label. In fact, compared to other sports drinks, Vita Coco contained far less sodium — an important electrolyte and mineral that is lost during sweating (Consumer Law Group, n.d.). Therefore, VitaCoco's products would not be as effective in replenishing electrolytes post physical exercise recovery as marketed (Consumer Law Group, n.d.).

Pickle Juice

In recent years, pickle juice has also gained traction for its potential health and exercise performance benefits by alleviating exercise-associated muscle cramps (EAMCs), which, according to commercial wisdom, are believed to be caused by electrolyte imbalances (Dreyfuss, 2016). The Pickle Juice Company, a popular pickle juice sports drink brand, markets its product as "10x-15x the electrolytes of common sports drinks" and "the only product scientifically proven to stop muscle cramps", insinuating that the electrolyte content of their pickle juice is what prevents muscle cramps (The Pickle Juice Company, n.d.-a). However, the scientific evidence presented by The Pickle Juice Company to consumers is limited and misrepresented. The brand's website only links a 2010 research study that reported pickle juice to help inhibit muscle cramping in hypohydrated individuals (Miller et al., 2010; The Pickle Juice Company, n.d.-b). Despite this finding, the study states that, "It is unknown which ingredient in pickle juice may initiate this inhibitory reflex. We propose that it is the acetic acid (vinegar) in pickle juice, not the electrolyte content, which triggers this reflex" — information that is omitted in their advertising (Miller et al., 2010).

In addition, the literature on muscle cramping being caused by electrolyte imbalances or pickle juice being effective at replenishing electrolytes is inconclusive at best (Dreyfuss, 2016). A 2014 study published in the Journal of Athletic Training investigated the effects of drinking pickle juice on muscle cramping in dehydrated individuals and found no appreciable changes in pickle juice's ability to restore the body's electrolyte or fluid losses (Miller, 2014). This finding is further supported by Schwellnus et al. (2011) and Shang et al. (2011) — both of which have found little evidence connecting electrolyte loss to EAMCs. The notion of pickle juice relieving muscle cramp is mainly derived from anecdotal evidence since the "electrolyte depletion" and "dehydration" hypotheses are unable to offer evidence-based pathophysiological mechanisms adequately explaining the presentation of EAMCs (Schwellnus, 2009). Regardless, The Pickle Juice Company continues to claim their products will cure EAMCs through electrolyte replenishment, which is a clear misrepresentation of scientific evidence.

Powerade
In 2019, two television advertisements marketing Powerade — which featured former rugby star, Israel Dagg and basketball player, Steven Adams — were taken off-air after complaints about their misleading narrative (Earley, 2019). The advertisements showed the athletes suffering an injury-related setback during their games, after which they drink Powerade and work hard until they triumphantly regain their original aptness in their respective sports (The New Zealand Herald, 2019). As these events take place, the voice-over in the advertisement states "Powerade ION4 replaces four electrolytes lost in sweat, to drive today's athletes forward" (The New Zealand Herald, 2019). The Advertising Standards Authority (ASA) ruled these advertisements as misleading, following a complaint that they insinuate Powerade to be "good for sports injuries". The complainant told the ASA that the advertisements "follow a very deliberate and consistent thread, and include the sports stars coming back from injury" and do not "talk about the injury, but the whole narrative is designed to imply that it,the sports drink, helps in such a scenario" (The New Zealand Herald, 2019). The complainant further told the ASA that their decision to file the complaint came after surveying a small cohort of children aged 10-13 on what the advertisement meant, where all children interpreted that Powerade would heal injuries incurred during sports (Earley, 2019). In response, Coca Cola argued that their advertisements were clear in only highlighting Powerade's ability to aid electrolyte loss during exercise without connecting it to athletes recovering from injury

(Earley, 2019). However, the ASA ruled in favour of the complainant, agreeing that the sequence of events portrayed in the advertisement were indeed deceptive (The New Zealand Herald, 2019).

Lucozade

A multimillion-pound advertising campaign for Lucozade Sport, a carbohydrate-electrolyte sports drink, was banned across the UK for false advertising (Shaikh, 2014). The advertisement shows two groups of men running on treadmills, one drinking water and the other drinking Lucozade Sport, as the voice-over says, "At the limits of your ability you need to replace the electrolytes you lose in sweat, keep your body hydrated, give your body fuel...Lucozade Sport gives you the electrolytes and carbohydrates you need, hydrating you, fuelling you better than water" (Cohen, 2014). The campaign resulted in 63 complaints being filed to the ASA for its potential breach of the advertising code (Cohen, 2014). One of these complaints was from the National Council of Hydration (also known as the UK Bottled Water Association) who insist that people who do not partake in intensive exercise actually have no need for sport drinks and regular water is well-sufficient for effective hydration (Cohen, 2014). In response, the Lucozade Sport's previous owner, GlaxoSmithKline (GSK), said that their health claims were consistent with the European Union legislation as well as the European Food Safety Authority (EFSA) following a scientific assessment conducted by them (Shaikh, 2014). GSK argued that the claim that Lucozade Sport provides better hydration that water did in fact completely align with two authorized claims by the EFSA's authorized that, "carbohydrate-electrolyte solutions enhance the absorption of water during physical exercise", and that such solutions aid with "the maintenance of endurance performance during prolonged endurance exercise" (Cohen, 2014). GSK further added that they followed the Department of Health's guidelines on flexibility in wording used in expressing health claims of a product (Cohen, 2014; Shaikh, 2014). Nevertheless, the ASA upheld their decision and stated that Lucozade Sport's advertising campaign was unclear in stating that the sports drink would only benefit individuals during periods of prolonged endurance exercise (Shaikh, 2014). The ASA also recommended that GSK should ensure that they are able to retain the original meaning of an authorized health claim, should they choose to re-word said health claim in order to promote customer understanding (Cohen, 2014). The National Council of Hydration rejoiced ASA's stance, commenting that ASA's decision to uphold the complaint would help clarify the widespread confusion among consumers on the actual role of sports drinks for the average person engaging in physical activity (BBC News, 2014).

There is also a layer of deficiency in scientific evidence that is important to discuss in this case as it relates sports drink advertising. Previously in 2012, the British Medical Journal (BMJ) raised objections to the evidence approved by the aforementioned EFSA, the agency of the European Union responsible for evaluating the scientific underpinnings of health or nutritional claims of sport-related supplements and drinks (Thompson et al., 2012). BMJ's findings questioned the premise that the surrogate physiological endpoint of hydration will induce significant differences in exercise performance (Cohen, 2014). According to the BMJ, the procedure used by the EFSA to evaluate the "nature and quality of the totality of the evidence" is not transparent or reproducible in several areas — and therefore, may not be considered scientifically rigorous (Thompson et al., 2012). Firstly, the EFSA relies on manufacturers of sports drinks to provide evidence speaking to the effectiveness or validity of their products. Hence, there is an opportunity for manufacturers to only present the EFSA with selective research that supports their product's health claims (Thompson et al., 2012). Secondly, the EFSA does not use any standard criteria to determine the type or quality of scientific evidence they will accept from manufacturers — opinion articles, book chapters and non-systematic review articles are all admissible to the EFSA along with scientific research studies (Thompson et al., 2012). Lastly, the BMJ was also unable to track EFSA's methodology to assess the scientific rigour of research studies, raising concerns on equal weighting being assigned by the EFSA to lower versus higher quality studies (Thompson et al., 2012).

Given the lack of a systematic and rigorous scientific review of evidence provided by sports drink manufacturers to food safety authorities, it begs the following question: Even if the advertised health claims about Lucozade Sport (or any other sport drink for that matter) are approved by authorities like the EFSA, are they truly authentic? Would such marketing be considered truthful?

Food for thought: The science of hydration and deceptive sports drink advertising

The BMJ has also conducted an extensive investigation on how the science of hydration is marketed by sports drink manufacturers (Cohen,

2012). According to the investigation, the greatest success of the sports drink industry was to convince the general public that the human body's natural homeostatic mechanism, i.e., thirst, is not an accurate system for responding to dehydration (Cohen, 2012). This convention is not actually scientifically supported — in fact, a meta-analysis of data of exercise-induced dehydration of cyclists in time trials suggested that fluid consumption relying on the natural instinct of thirst was the best hydration strategy (Cohen, 2012).

Regardless, sports drink companies perpetuated the unreliability of the human thirst mechanism by sponsoring scientists who eventually developed a new field of science devoted to hydration (Cohen, 2012). These scientists then collaborated with leading sports medicine organizations to develop hydration guidelines that have gone to influence nutrition/sport advice from Olympic Committees, the EFSA as well as every day healthcare organizations (Cohen, 2012). In doing so, these companies and scientists have inspired great concern about the dangers of dehydration — a public concern that conveniently furthers their financial interests by allowing them to sell more products (Cohen, 2012).

One of the dangers of dehydration that has been sold to consumer is hyponatremia (i.e., insufficient levels of the sodium electrolyte) and that exercise-associated hyponatremia is best prevented by a sports drink rather than regular water, for athletes and non-athletes alike (Cohen, 2012). For instance, The Coca-Cola Company suggests that hyponatremia is a cause for concern "for anyone doing endurance sports" and is caused by the failure to "replace the sodium lost through sweat or drinking a very large volume of very low-sodium beverages such as water"(Cohen, 2012). However, exercise-induced hyponatremia is primarily a concern for endurance athletes that engage in high-intensity exercise for a prolonged time period, not for most members of the general population that engage in regular exercise — they do not exercise as long or as intensely to require sports drinks (Tinsely, 2018). In fact, there is general consensus that water can hydrate most active people just as effectively as sports drinks and they do not need to consume fluids beyond water to replenish electrolytes (Tinsely, 2018; Tuller, 2012).

The bottom line is that deceptive sports drink advertising has shifted consumer perceptions on the importance of hydration in daily life and the types of fluid to hydrate oneself with, ultimately to protect their financial interests.

CHAPTER 10

Electrolyte Drink Marketing: Professional Athletes

Amna Abu Askar

While electrolyte beverages, more commonly known as sports drinks, are freely available on the market, intense marketing efforts are geared towards professional athletes and individuals engaged in regular physical activity (Market Data Forecast, 2020). The global electrolyte drinks market size has increased over the years with an estimated value of 1.42 billion USD in 2021, and is predicted to grow to 1.82 billion USD by 2025 (Market Data Forecast, 2020). North America and Europe are the largest electrolyte drink markets, representing about 50% of the total share (Market Data Forecast, 2020). The rise in market size is due to the increased preference for healthier beverages over sugar-based drinks before, during, or after exercise completion. Furthermore, a booming interest in health and fitness amongst teens and adults has increased the consumer base for the market, leading to market size growth. Hence, manufacturers are continuously introducing different innovative flavours to appeal to different age groups and improve palatability (Market Data Forecast, 2020).

Electrolyte drinks are often confused with energy drinks. Unlike electrolyte drinks, energy drinks contain stimulants such as caffeine, taurine, and guarana and are marketed for enhancing mental alertness and for physical stimulation (Wyckoff, 2011). While there are many benefits associated with electrolyte drinks, especially to those involved in vigorous physical activity, energy drinks with high concentrations of caffeine can affect the appropriate balance of macromolecule (carbohydrates, fat, and protein) intakes. Too much caffeine can also increase blood pressure, heart rate, and trigger irregular heart rhythm (Wyckoff, 2011). It can lead to sleeplessness, headaches, anxiety and concentration difficulties (Callahan, 2014). According to a study by American Academy of Pediatrics (APP), some energy drinks contain more than 500mg of caffeine, a value nearing caffeine toxicity levels

(Wyckoff, 2011). As a result, energy drinks are not recommended to children or adolescents, as it can negatively affect proper growth and development; specifically the development of their cardiovascular and nervous systems (Wyckoff, 2011). Most concerning is the lack of safety regulations of these energy drinks and the aggressive marketing towards adolescents (Al-Shaar et al, 2017). Moreover, regardless of whether energy drinks are being marketed as beverages or dietary supplements, there is no requirement to declare on the label the amount of caffeine present, which makes it difficult to identify the actual caffeine content present in energy drinks (NCCIH, 2018). It is also worth noting that a single 16oz energy drink bottle can contain around 54 to 62 grams of added sugar, which exceeds the maximum amount recommended for a day (NCCIH, 2018). Although electrolyte drinks' composition is similar, electrolyte drinks do not contain stimulants nor are they packed with added sugar.

Generally, electrolyte drinks help in recovery and energy maintenance after strenuous training and are advertised as such. These drinks help prevent dehydration and replenish the body with vital electrolytes including sodium, potassium, magnesium, calcium, and phosphate, lost through excessive sweating and which are important for stimulating muscles and nerves (Market Data Forecast, 2020). Sport drinks also contain 6-8% carbohydrates in the form of simple sugars such as glucose, fructose, and sucrose for quick energy replenishment. Some drinks, however, are low or zero-carbs to appeal to consumers looking to obtain electrolytes and water without additional calories. Although several brands exist like Gatorade, Powerade, and AllSport, no difference in effectiveness was found across the different sport beverages available on the market (Coombes & Hamilton, 2000). In addition to water, carbohydrates, and electrolytes, sport drinks can contain omega-3 fatty acids for cardiovascular health, fibre and probiotics to help improve gut motility and weight management, collagen to help enhance skin appearance, and vitamin D and zinc to strengthen immunity (Orrù et al., 2018). Various amino acids can help reduce fatigue and enhance muscle function, B vitamins can also be added to boost metabolism (Orrù et al., 2018). Other functional ingredients often found in these electrolyte drinks include stabilizers, flavors, sweeteners, and colors (Orrù et al., 2018). Hence, sport drinks are often advertised for supporting the immune system, improving gut and cardiovascular health, or counteracting the aging process (Kenefick & Cheuvront, 2012). While these are all favourables outcomes, the primary feature that drives the success of the electrolyte drinks' market is that it is a convenient fuel in terms of size, storage, and providing the desired nutrients and bioactive compounds (Wootton-Beard & Ryan, 2011).

Many might argue, however, that these electrolyte drinks are unnecessary and water suffices hydration needs. While this might be true for an untrained person, athletes sweat much more than the average person because their bodies have adapted to remove heat more efficiently, losing more water and electrolytes (Nelson, 2021). Dehydration and electrolyte imbalance reduces an athlete's training capacity and performance. Thus, if athletes were to simply rehydrate with water, they could further throw off body balance by diluting their electrolyte concentrations (Fischer-Colbrie, 2017). Inadequate electrolyte levels can lead to dangerous conditions such as hyponatremia (low sodium levels), which can lead to seizures, coma and even death if not treated immediately (Hyponatremia, 2015). Hence, electrolyte drinks can help optimize athletic performance, reduce fatigue and health conditions associated with dehydration and electrolyte imbalancement in a convenient way, while reducing stomach upsets and maximizing intestinal absorption and energy delivery to the muscles.

Various clinical trials were conducted to determine how relevant electrolyte drinks are for athletes and how much added benefit they provided. One study compared the performance of trained cyclists who drank carbohydrate-electrolyte beverages to those who received the placebo drinks - artificially flavored and colored water - during a 1-hour intense cycling session (Jeukendrup et al., 1997). They found that carbohydrate-electrolyte drinks enhanced performance by 2% compared to the placebo. While a 2% increase in performance might seem a negligible benefit to the average person, these results are absolutely crucial for a competitive athlete. The success of an athlete's competitive performance is often determined by the narrowest of margins. For example, during the women's triathlon of the 2012 summer olympics in London UK, Nicola Spirig finished first place with a total time of 1:59:48, while Ainhoa Murua ranked 7th place with a finishing time of 2:00:56 (IOC, n.d.). The difference between securing the olympic champion title and coming in 7th place was 0.9%. Hence, any supplement or sports drink that can improve performance by even 1% is crucial for a professional athlete. Another report examined nine studies of high intensity cycling or running lasting a short period of 30-60 minutes. Of the nine studies investigated, six demonstrated that ingesting sports drinks before or during exercise had a beneficial effect on the performance of elite athletes (Coombes & Hamilton, 2000). Despite these findings, there is little evidence to suggest that electrolyte drinks can enhance performance of short-duration based activities such as sprinting, agility, and jumping exercises (Baker et al., 2015). Furthermore, it has been shown that water and electrolytes lost during short-duration exercise does not limit exercise performance. Hence, the need for fluid and electrolyte replacement is thought to be negligible during intense exercises that last for an hour or less (Coombes & Hamilton, 2000).

However, in team sports or intermittent exercises such as soccer and rugby, where players alternate between intense exercise and rest, sport drinks are shown to reduce fatigue and improve performance compared to a placebo (Baker et al., 2015). One report examined 12 studies designed to mimic the intermittent nature of many team sports by using prolonged intermittent exercise followed by brief sprints, and found that 9 studies showed that sport drinks significantly enhanced performance compared to placebo (Coombes & Hamilton, 2000). Although many aspects of intermittent exercises are carbohydrate dependent include high intensity efforts, aerobic activity, cognitive demands related to skill, attention, and decision-making, 7 of the 9 studies used low-carb drinks (10% carbohydrates or less) (Coombes & Hamilton, 2000). Another study conducted in India, tested the effect carbohydrate-electrolyte drinks have on performance, blood glucose and lactate levels in 16 male endurance athletes (Brouns et al., 1998). The athletes were to run on a treadmill at 70% of their maximum aerobic capacity, and their heart rate was monitored. Brouns et al. (1998) found that the total endurance time at 70% VO2 max and cardiovascular responses significantly improved in athletes who consumed carbohydrate-electrolyte drinks compared to those who did not. This suggests that carbohydrate and electrolyte balance keeps heart rate and lactate levels low during exercise and delays fatigue onset. It was also noted that blood glucose and peak lactate levels during exercise and recovery were the same whether athletes were supplemented with carbohydrate-electrolyte drinks or not. However, the removal of blood lactate during recovery was faster after 10 and 20 minutes of supplementing with 12.5g per cent carbohydrate-electrolyte beverage, thereby delaying tiredness and fatigue. Likewise, many studies examining the effects of sports drinks on athletes involved in long-duration continuous exercises lasting between 1-4 hours or longer, such as endurance sports, showed major improvements in performance (Jeukendrup, 2011). With increases in exercise duration without rest, the number of carbohydrates recommended to maintain exercise intensity increases from 30g per hour to 60g per hour, however, these recommendations do not apply to weight training sessions (Jeukendrup, 2011). Carbohydrate and electrolyte requirements also vary from person to person based on age, sex, exercise duration, and other environmental factors (Market Data Forecast, 2020).

Not only are electrolyte drinks used by elite athletes during competitions and training sessions to replace fluids and electrolytes, but many electrolyte drinks brands are promoted by sports champions. For example, the "Bolt Lemon Ice" by Gatorade was inspired by Usain Bolt, the world's fastest man, and 100m and 200m record holder. The drink contains 6% carbohydrates and helps in quick refueling of the working muscles. In fact, the drink is composed of the same formula as the other Gatorade drinks with the only difference being the new lemon ice flavour. The "Bolt Lemon Ice" drink has been available since August 2010 in Australia and is being largely endorsed by the Australian Institute of Sport, Sports Dietitians Australia, and Victorian Institute of Sport (Drink Like A Champion - Celebrity Drink Endorsements, 2008). With Gatorade dominating the sport drinks industry, Powerade has also been relying on celebrity endorsement to promote its products (Horrigan, 2016). Some of the top athletes who have endorsed Powerade include Derrick Rose and LeBron James (Horrigan, 2016). Based on 2010 data from an advertisement database, Nielson and AdScope, commercials of foods and beverages endorsed by professional athletes were mainly viewed by adolescents between the ages 12 and 17. Thus, athlete endorsements highly influence adolescents' consumption attitudes and food choices.

The marketing of sports drinks and other sports products has become a multibillion dollar industry. According to the 2010 National Health Interview Survey, 31.3% of adults were regular sport drink consumers, with 11.5% consuming these drinks 3-4 times per week. Surprisingly, a large number of children and adolescents also consume sports drinks outside of sporting activities (Broughton et al., 2016). With this overconsumption of sports drinks comes an increased risk of dental caries and erosion, since sports drinks contain acids and simple sugars (Broughton et al., 2016). Furthermore, if electrolyte drinks are to be consumed socially and in excessive quantities, serious health conditions such as obesity, diabetes, heart disease and gout can arise (Broughton et al., 2016). Studies have shown that most non-athletes consume electrolyte beverages simply because they taste good (Healthy Eating

Research, 2012). Although evidence suggests that sport drinks deliver ergogenic effects to elite athletes involved in intense sporting events, there is no evidence of beneficial effects on non-athletes or children.

In summary, water, electrolytes, and carbohydrates are critical nutrients for normal physiological functioning and for optimal athletic performance. As exercise intensity increases, fluid and electrolytes are lost, unless compensated for with appropriate intakes. While drinking water and eating a balanced meal before or during training sessions and competitions can help restore energy levels, sports drinks conveniently deliver all the nutrients and electrolytes needed to keep the body going, reduce fatigue during prolonged physical activities, and aid performance, in a single bottle. Electrolyte drinks are an exponentially growing segment of beverages on the market. With a consumer base that has expanded from elite athletes and endurance runners to non-elite athletes, adolescents and children. The growth in market size is largely attributed to increased interest in fitness and healthy living, however, athlete endorsement of sports drinks and sport products has a profound effect on adolescents' consumption habits.

CHAPTER 11

The Future of Electrolyte Drinks

Ami Patel

At present, electrolyte drinks are typically used in very niche situations, such as by athletes for exercise or by children to prevent dehydration (through products such as Pedialyte). But with all the previously mentioned functions of electrolytes in the human body, electrolyte-containing drinks may have much broader future uses in various situations and populations. In the future, electrolyte drinks may potentially be available to enhance cognitive performance during and after exercise (and potentially outside of athletics too). They may help treat symptoms in patients presenting with various mental illnesses. They could be used to address dehydration globally, including dehydration as a result of diarrhea. Lastly, they may be beneficial for other miscellaneous events such as treating a hangover. This chapter will discuss the rationale behind the potential future applications of electrolyte drinks in more detail.

Electrolyte Drinks and Cognitive Performance
High-intensity exercise (such as running) can induce dehydration and heat stress, especially when performed in a hot and humid environment (Wong et al., 2014). A consequence of heat stress and dehydration is an alteration to the central nervous system (CNS), which can manifest through central fatigue (lower muscle force due to lower output from motor neurons; Gandevia et al., 1995) and limit an individual's performance in the heat (Wong et al., 2014). Dehydration and heat stress can also cause significant reduction in other cognitive functions such as by impairing perceptive discrimination, short-term memory (also known as working memory), an individual's subjective estimates of fatigue, math performance, and visuomotor tracking (Wong et al., 2014). These negative consequences of dehydration and heat stress can impact cognitive functions in other populations as well, such as children studying in countries with extreme heat and inadequate access to water.

In Wong et al. 's (2014) study, to help individuals rehydrate during exercise, they provided carbohydrate-electrolyte solutions that help replace the electrolytes lost while sweating. Next, they measured their performance on a working memory task (Wong et al., 2014). Individuals that consumed a carbohydrate-electrolyte solution or lemon tea to rehydrate performed with greater accuracy on the working memory task compared to those that had distilled water. The authors suggest that during a short-term recovery (a few hours after exercise), both carbohydrate-electrolyte solution and lemon tea are advantageous for maintaining cognitive performance (or recovering lost cognitive performance due to dehydration or heat stress). Thus, future electrolyte drinks designed for athletes may contain more cognition-enhancing features to improve and maintain an athlete's performance during exercise. The idea of cognitive enhancement raises ethical concerns, as such drinks, if available, could be abused by athletes to enhance performance and gain an unfair advantage over other participants.

Another way to improve cognitive function is by combining an electrolyte drink with another substance or chemical. Pruna et al. (2014) performed a study examining the effects of rehydrating athletes performing endurance exercise by combining alanine-glutamine dipeptide with a commercially available sports drink (contains electrolytes). The results suggest that combining these two substances provided greater ergogenic benefits (improvements in physical performance, stamina and recovery) to athletes compared to benefits from consuming only a commercial sports drink. Cognitive function was also suggested to improve, demonstrated through enhanced reactivity to multiple visual stimuli. Based on these results, the authors suggest the additional benefit provided by alanine-glutamine dipeptide may be due to its ability to improve fluid and electrolyte absorption in the stomach. The enhanced absorption may allow athletes to maintain fine motor control and reactivity throughout their exercise regimen.

In the future, electrolytes may contain substances like alanine-glutamine dipeptide to enhance athletes' cognitive performance during exercise. If such a drink is made available, the possibility of a super drink containing electrolytes, substances that increase electrolyte absorption, and other substances such as caffeine may not be too far away. A super drink could be used as a meta-enhancement for physical and mental performance and may not be restricted to use only by athletes. If a super drink is made available to improve cognitive performance in other domains, it has wide uses in the population. A specific example is workers and students consuming these drinks to improve productivity.

Applications to Mental Health & Illnesses

Making Exercise More Pleasurable

An estimated 45% of those that begin an exercise program drop out (Marcus et al., 2006). Electrolyte drinks in the future may be able to help individuals turn exercising regularly into a habit. Having a pleasurable affective response (pleasurable moods and feelings) during exercise and thus, forming a positive memory of the exercise experience may directly increase the odds of adhering to a long-term exercise program (Peacock et al., 2012). A few studies have found that when individuals drink water or a carbohydrate-electrolyte drink during exercise, they report feeling better (Peacock et al., 2012). Such positive affective states may play a role in how long individuals persist with an exercise routine and the amount of effort they put into performing exercises (Peacock et al., 2012). Individuals that consumed carbohydrate-electrolyte solutions during exercise had better positive affective states (rated on a pleasure-displeasure scale) compared to individuals that consumed water (Peacock et al., 2012). Thus, in addition to widely known physiological benefits, electrolyte drinks may also have psychological benefits that help individuals maintain an active lifestyle (Peacock et al., 2012). Such maintenance is possible through positive experiences with exercise that result in greater enjoyment of the activity; individuals that enjoy an activity may be more likely to persist with it long-term to reap the rewards (Peacock et al., 2012). More pleasurable responses to exercises may also be associated with a higher amount of time allocated to daily exercise (Peacock et al., 2012). Evidently, electrolyte drinks may create more positive experiences with exercise, encouraging individuals to turn exercising into a regular activity.

Mental Illness-Induced Polydipsia

An estimated 10%-25% of those diagnosed with chronic schizophrenia develop polydipsia, defined as excessive drinking or thirst (Quitkin et al., 2003). A third of these patients become hyponatremic (having low sodium concentration in the blood; refer to Chapter 3 for more details) (Quitkin et al., 2003). If sodium concentrations fall below 120 mmol/litre, seizures, coma, and death may occur (Quitkin et al., 2003). In a single case, a patient suffered two life-threatening episodes of coma due to hyponatremia because they were unable to stop consuming fluids in excess (Quitkin et al., 2003). After behavioural and pharmacological treatments were unsuccessful in treating the patient's polydipsia, the patient was advised to drink only an electrolyte sports drink and take a salt pill with every meal (salt contains sodium, an electrolyte) (Quitkin et al., 2003). After implementing this recommendation, the patient demonstrated improvements in mental status through enhanced orientation and stable sodium concentrations (Quitkin et al., 2003). This example demonstrates a potential future application of electrolyte drinks for addressing polydipsia in patients with mental illnesses like schizophrenia, anxiety disorders, mood disorders, personality disorders, and eating disorders such as anorexia nervosa (D'Arrigo, 2017). However, it is important to note that the above example was a single case study. Thus, additional studies must replicate these findings using appropriate scientific methods before any consideration of implementation.

Electrolyte Imbalance Disorders
There are various electrolyte imbalance-related disorders. These include electrolyte disorders due to imbalances in sodium (hypernatremia and hyponatremia), potassium (hyperkalemia and hypokalemia), calcium (hypercalcemia and hypocalcemia), chloride (hyperchloremia and hypochloremia), phosphate (hyperphosphatemia or hypophosphatemia), or magnesium (hypermagnesemia and hypomagnesemia) (Holland, 2013). Descriptions of some of these specific conditions are available in more detail in previous chapters. Conditions such as hyponatremia, hypochloremia, hypomagnesemia, hypophosphatemia, and hypokalemia may be a result of an underlying mental illness, which includes alcohol use disorders and eating disorders (Holland, 2013). However, all of the listed electrolyte disorders can be caused by other physical disorders as well. Overall, any of the above-mentioned conditions (especially those that concern the lack of a certain electrolyte) may in the future involve electrolyte drinks as a potential treatment method. Please note that there is little scientific research as of right now that explores such alternative treatments for most of these conditions.

Applications to Global Health
As mentioned in Chapter 5, oral rehydration therapy (ORT), such as commercially available Pedialyte electrolyte drinks for children, is used to treat dehydration resulting from diarrhea in children. ORT is the preferred treatment for dehydration as it functions to restore circulating blood volume, restore interstitial fluid volume and maintain hydration (Canavan & Arant, 2009). ORT is also more advantageous than intravenous fluid therapy to treat dehydration (Canavan &

Arant, 2009). A specific advantage is that ORT can be administered at home (children can consume Pedialyte solutions at home under the supervision of their parents; does not require medical professionals or a visit to a medical facility/emergency department) (Canavan & Arant, 2009). Other benefits include that the same ORT fluid had multiple uses (rehydration, maintenance, and replacement of stool lost due to diarrhea) and treatment with ORT can be provided faster than intravenous fluid therapy (Canavan & Arant, 2009). Typically, commercial ORT solutions such as Pedialyte are used in the Western world to correct electrolyte imbalances due to gastroenteritis in children (Canavan & Arant, 2009). Globally, such treatments have other applications and may potentially have even wider applications in the future.

More than 2 billion people worldwide drink water from fecally contaminated water sources (World Health Organization, 2018). Drinking contaminated water can cause cholera, an infectious disease that causes diarrhea (World Health Organization, 2021). If left untreated, cholera can cause death within hours (World Health Organization, 2021). According to the World Health Organization (2021), an estimated 1.3 to 4 million cases of cholera occur each year, causing 21 000 to 143 000 deaths annually. Fortunately, most of those that catch an infection can be treated with an oral rehydration solution (ORS), similar to ORT. With treatment, 99% of infected individuals survive (World Health Organization, 2018; World Health Organization, 2021). The World Health Organization's ORT solution for cholera contains enough sodium to replace the amount typically lost due to diarrhea (Canavan & Arant, 2009). By 2030, the Global Task Force on Cholera Control (GTFCC) has a goal to reduce the deaths caused due to cholera by 90% (World Health Organization, 2021). Electrolyte drinks may play a major role in helping to reduce cholera infections and deaths and potentially eliminating the disease in some countries.

In addition to treating cholera, electrolyte drinks can also be used to treat severe malnutrition in infants and children. Severe acute (short-term) malnutrition is a major cause of child mortality worldwide, with severe wasting accounting for an estimated 400 000 child deaths annually (World Health Organization, 2013). Children typically develop malnutrition when they are around six to eighteen months old, as they are growing and their brain is developing rapidly (World Health Organization, 2013). This means young children and infants are highly susceptible to malnutrition if their diets do not contain adequate nutrient content, or their food is contaminated (which may also cause diseases like cholera) (World Health Organization, 2013).

Acute malnutrition and diarrhea also have a bidirectional relationship: acute malnutrition increases a child's chances of getting diarrhea and the duration of diarrhea, while diarrhea can cause or worsen malnutrition (World Health Organization, 2013). Diarrhea decreases absorption of carbohydrates, proteins, other nutrients, and electrolytes like potassium, which contributes to malnutrition (World Health Organization, 2013). Diarrhea also causes high water loss, which can lead to dehydration, electrolyte imbalance, shock and eventually death (World Health Organization, 2013).

Children that have severe acute malnutrition and moderate to severe dehydration can be rehydrated with oral rehydration solution (ORS) (World Health Organization, 2013). Children can also be adequately nourished with therapeutic food, which contains many essential nutrients (vitamins, zinc, micronutrients, etc.) and electrolytes to address deficiencies caused by malnutrition (World Health Organization, 2013). Thus, electrolyte drinks may be used to complement other electrolyte providing sources (food) in cases of severe electrolyte imbalance and malnourishment in children.

The desired effect of fluid management of children with severe acute malnutrition includes reduced mortality, avoidance of electrolyte abnormalities, and reduced duration of diarrhea (World Health Organization, 2013). Due to the high number of child deaths caused by malnourishment, the World Health Resolution on Infant and Young Child Nutrition is working to improve their approach to managing severe acute malnutrition (World Health Organization, 2013). Future electrolyte drinks may be able to help achieve this goal by improving the contents inside the drinks and catering them specifically to malnourished children. This could help create more efficient products, allowing the WHO to tackle global childhood malnutrition more effectively.

Other Applications: Alcohol Consumption
An estimated one-third of US citizens consume alcohol, costing the American economy $148 billion annually in extra sick days and lost productivity due to poor work performance (Wiese et al., 2000). Alcohol is the most commonly used drug in the world, yet there is no medical consensus on the best methods for treating hangovers (Jin,

2019). Hangovers present with different symptoms for different people; some symptoms include confusion, headaches, shakiness, nausea, vomiting, diarrhea, and fatigue (Jin, 2019). So what causes these symptoms? Ethanol (also called alcohol) affects various physiological processes. One example is the antidiuretic hormone (ADH), which regulates volume and electrolytes through the kidneys, preventing urination (Jin, 2019). When a person consumes ethanol, their ADH is inhibited, increasing the individual's frequency of urination and the water concentration in the urine (Jin, 2019). The result is dehydration and loss of vitamins, proteins, and electrolytes (like sodium and potassium) (Jin, 2019).

To combat the consequences of alcohol use and possibly even prevent the intensity of hangover symptoms the following morning, some individuals consume Pedialyte (Jin, 2019). In this situation, Pedialyte may help maintain electrolyte balance while drinking, potentially reducing the symptoms of alcohol consumption. Since Pedialyte helps with rehydration, it may also prevent dehydration, which would decrease the likelihood of experiencing headaches and fatigue (Jin, 2019). Consuming Pedialyte in the morning after drinking could also help continually address dehydration symptoms (Jin, 2019). Additionally, drinking Pedialyte could help combat the negative consequences of vomiting and diarrhea on maintaining electrolyte balance. Thus, electrolyte drinks may help combat the overall negative effects of consuming alcohol. Since Pedialyte is a drink made for children, other flavoured drinks could be created to specifically target adults that are drinking. Such drinks could be consumed alongside alcohol, mixed into alcohol (if research shows that it is beneficial), or consumed the day after drinking, to deal with hangover symptoms. If such an electrolyte drink could be created to help reduce consequences in various stages of alcohol consumption, it may increase alcohol consumption as a side effect. This topic brings about ethical considerations that must be acknowledged before the development and marketing of such a substance.

Conclusion
In conclusion, electrolyte drinks currently only have a select few applications, such as sports drinks for athletes or rehydration solutions for children. At present, the primary goal of these solutions is to restore hydration and electrolytes lost due to sweat or other conditions like diarrhea. In more recent years, electrolyte drinks are starting to be used in other settings, such as reducing symptoms that result from

mental illness or treating severe malnourishment in children globally. Electrolyte drinks have much more potential and may soon be used in broader settings. Future electrolyte drinks may play a role in anything ranging from addressing conditions causing physiological imbalances to enhancing thinking and regulating psychological affect. With so many technological advances, we may be well on our way towards improving cognition and addressing the consequences of global dehydration. Once firm scientific research confirms or disproves the varying suggested applications of electrolyte drinks, we may not be too far from having access to all-in-one super drinks. Thus, several questions arise. Who would consume these super drinks? Would you? Are these drinks ethical? Or should researchers and corporations prioritize making electrolyte drinks for certain groups (starving and dying children) before super drinks?

REFERENCES

Chapter 1.

Adar, R. M., Markovich, T., & Andelman, D. (2017, May 17). Bjerrum pairs in ionic solutions: A poisson-boltzmann approach. *Journal of Chemical Physics, 146(*19404). https://doi.org/10.1063/1.4982885

Bergersen, B. (1998, September 14). *The historical origin of statistical mechanics.* Retrieved from UBC Department of Physics & Astronomy: https://phas.ubc.ca/~birger/boltzmann/node2.html

Bjerrum, N. (1909, May-June 27-3). *A new form for the electrolytic dissociation theory.* Retrieved from ChemTeam: https://www.chemteam.info/Chem-History/Bjerrum-Strong-Electrolyte.html

Boeck, G. (2019, January 16). *Paul walden (1863-1957): The man behind the walden inversion, the walden rule, the ostwald-walden-bredig rule and ionic liquids.* Retrieved from ResearchGate: https://www.researchgate.net/publication/330759696_Paul_Walden_1863-1957_the_man_behind_the_Walden_inversion_the_Walden_rule_the_Ostwald-Walden-Bredig_rule_and_Ionic_liquids

Bradley, S., Kammermeier, K., & Chemistry LibreTexts. (2020, September 23). *Debye-hückel theory.* Retrieved from Chemistry LibreTexts:https://chem.libretexts.org/Bookshelves/Physical_and_Theoretical_Chemistry_Textbook_Maps/Supplemental_Modules_(Physical_and_Theoretical_Chemistry)/Physical_Properties_of_Matter/Solutions_and_Mixtures/Nonideal_Solutions/Debye-H%C3%BCckel

De Berg, K. C. (2003). The development of the theory of electrolytic dissociation. *Science & Education, 12,* 397-419. https://doi.org/10.1023/A:1024438216974

Ebeling, W., & Sokolov, I. (2005). *Statistical thermodynamics and stochastic theory of nonequilibrium systems.* World Scientific Publishing Company. DOI:10.1142/2012

Faraday, M., & Royal Society (Great Britain). (1833, February 19). *On electro-chemical decomposition.* The Royal Society. Retrieved from

Smithsonian Libraries: https://library.si.edu/digital-library/book/onelectrochemica00fara

Hall, N. F. (1929, May 1). Salts, acids, and bases: Electrolytes: Stereochemistry (walden, paul). *Journal of Chemical Education, 6(5),* 1005. https://doi.org/10.1021/ed006p1005

Huebener, R. P. (2013, January). *Walther nernst: Physicist and chemist with great vision.* Retrieved from The Electrochemical Society: https://knowledge.electrochem.org/encycl/art-n02-nernst.htm

Jones, H. C. (1899). *The modern theory of solution: Memoirs by pfeffer, van't hoff, arrhenius, and raoult.* American Book Company.

Jorpes, J. E. (1970). *Jac. berzelius: His life and work.* California: University of California Press. https://doi.org/10.1021/ed043pA1100.1

Kormos-Buchwald, D. L. (2020, November 24). *Walther nernst.* Retrieved from Encyclopedia Britannica: https://www.britannica.com/biography/Walther-Nernst

Kunz, W. (2014, September 25). Electrolytes, history. (G. Kreysa, K.-i. Ota, & R. F. Savinell, Eds.) *Encyclopedia of Applied Electrochemistry.* https://doi.org/10.1007/978-1-4419-6996-5

LeTran, D., & Chemistry LibreTexts. (2020, August 15). *The ideal gas law.* Retrieved from Chemistry LibreTexts: https://chem.libretexts.org/Bookshelves/Physical_and_Theoretical_Chemistry_Textbook_Maps/Supplemental_Modules_(Physical_and_Theoretical_Chemistry)/Physical_Properties_of_Matter/States_of_Matter/Properties_of_Gases/Gas_Laws/The_Ideal_Gas_Law

Malley, K., Artika, S., & Chemistry LibreTexts. (2020, September 1). *Ionic activity.* Retrieved from Chemistry LibreTexts: https://chem.libretexts.org/Bookshelves/Physical_and_Theoretical_Chemistry_Textbook_Maps/Map%3A_Physical_Chemistry_for_the_Biosciences_(Chang)/05%3A_Solutions/5.08%3A_Ionic_Activity

Mander, P. (2017, February 1). *Carlisle, nicholson and the discovery of electrolysis.* Retrieved from Carnotcycle - the classical blog on thermodynamics: https://carnotcycle.wordpress.com/2017/02/01/carlisle-nicholson-and-the-discovery-of-electrolysis/

MIT Libraries. (2021). *The voltaic pile*. Retrieved from MIT Libraries - Distinctive Collections Spotlights: https://libraries.mit.edu/collections/vail-collection/topics/electricity/the-voltaic-pile/

Nobel Media AB. (2021). *Walther nernst biographical*. Retrieved from The Nobel Prize: https://www.nobelprize.org/prizes/chemistry/1920/nernst/biographical/

Onsager, L., & Fuoss, R. M. (1932, November 1). Irreversibly processes in electrolytes. diffusion, conductance, and viscous flow in arbitrary mixtures of strong electrolytes. *Journal of Physical Chemistry, 36*(11), 2689-2778. https://doi.org/10.1021/j150341a001

Oxford Reference. (2021). *Debye-hückel theory*. Retrieved from Oxford Reference: https://www.oxfordreference.com/view/10.1093/oi/authority.20110803095705261#:~:text=A%20theory%20to%20explain%20the,electrostatic%20interactions%20between%20the%20ions.

Oxford Reference. (2021). *Walden's rule*. Retrieved from Oxford Reference: https://www.oxfordreference.com/view/10.1093/oi/authority.20110803120408835#:~:text=An%20empirical%20rule%20suggested%20by,same%20ions%20in%20different%20solvents.

Parker, S. (2017, May 9). *Osmotic investigations: Studies on cell mechanics (1877), by wilhelm pfeffer*. Retrieved from The Embryo Project Encyclopedia: https://embryo.asu.edu/pages/osmotic-investigations-studies-cell-mechanics-1877-wilhelm-pfeffer

Ramberg, P. J. (2021, February 25). *Jacobus henricus van't hoff*. Retrieved from Encyclopedia Britannica: https://www.britannica.com/biography/Jacobus-Henricus-van-t-Hoff

Schummer, J. (2021, March 31). *Wilhelm ostwald*. Retrieved from Encyclopedia Britannica: https://www.britannica.com/biography/Wilhelm-Ostwald

Science History Institute. (2017, December 4). *Humphry davy*. Retrieved from Science History Institute: https://www.sciencehistory.org/historical-profile/humphry-davy

Science History Institute. (2017, December 1). *Jöns jakob berzelius*. Retrieved from Science History Institute: https://www.sciencehistory. org/historical-profile/jons-jakob-berzelius

Smith, R., Inomata, H., & Peters, C. (2013, December 12). Chapter 6 - equations of state and formulations for mixtures. *Supercritical Fluid Science and Technology, 4,* 333-480. https://doi.org/10.1016/B978-0-444-52215-3.00006-4

Speight, J. (2021). *Natural water remediation.* Cambridge: Butterworth-Heinemann. https://doi.org/10.1016/C2013-0-16022-9

Surat, P. (2019, January 9). *History of electrochemistry.* Retrieved from News Medical: https://www.news-medical.net/life-sciences/History-of-Electrochemistry.aspx#:~:text=One%20of%20the%20first%20 experiments,next%20to%20an%20electrostatic%20generator.

The Editors of Encyclopaedia Britannica. (2011, February 8). *Statistical mechanics.* Retrieved from Encyclopedia Britannica: https://www. britannica.com/science/statistical-mechanics

The Editors of Encyclopaedia Britannica. (2017, June 7). *Electrolyte.* Retrieved from Encyclopedia Britannica: https://www.britannica.com/ science/electrolyte.

The Editors of Encyclopaedia Britannica. (2021, March 25). *Faraday's laws of electrolysis.* Retrieved from Encyclopedia Britannica: https:// www.britannica.com/science/Faradays-laws-of-electrolysis

The Editors of Encyclopaedia Britannica. (2021, March 5). *Wilhelm pfeffer.* Retrieved from Encyclopedia Britannica: https://www. britannica.com/biography/Wilhelm-Pfeffer

The Electrochemical Society. (2021). *Birth of electrochemistry.* Retrieved from The Electrochemical Society: https://www.electrochem.org/ birth-of-electrochemistry

Chapter 2.

Adrogué, H. J., & Madias, N. E. (2007). *Sodium and Potassium in the Pathogenesis of Hypertension | NEJM*. New England Journal of Medicine. https://www.nejm.org/doi/10.1056/NEJMra064486

Bennett, J., Deslippe, A. L., Crosby, C., Belles, S., & Banna, J. (2020). Electrolytes and Cardiovascular Disease Risk. *American Journal of Lifestyle Medicine, 14*(4), 361–365. https://doi.org/10.1177/1559827620915708

Bilaloglu, S., Aphinyanaphongs, Y., Jones, S., Iturrate, E., Hochman, J., & Berger, J. S. (2020). Thrombosis in Hospitalized Patients With COVID-19 in a New York City Health System. *JAMA, 324*(8), 799–801. https://doi.org/10.1001/jama.2020.13372

Bradshaw, B. G., Liu, S. S., & Thirlby, R. C. (1998). Standardized Perioperative Care Protocols and Reduced Length of Stay After Colon Surgery. *Journal of the American College of Surgeons, 186*(5), 501–506. https://doi.org/10.1016/S1072-7515(98)00078-7

Carriazo, S., Kanbay, M., & Ortiz, A. (2020). Kidney disease and electrolytes in COVID-19: More than meets the eye. *Clinical Kidney Journal, 13*(3), 274–280. https://doi.org/10.1093/ckj/sfaa112

Chowdhury, A. H., & Lobo, D. N. (2015). Water and Electrolytes. In *Clinical Nutrition* (pp. 27–46). John Wiley & Sons, Ltd. https://doi.org/10.1002/9781119211945.ch3

Chowdhury, A. H., & Lobo, D. N. (2015). Water and Electrolytes. In *Clinical Nutrition* (pp. 27–46). John Wiley & Sons, Ltd. https://doi.org/10.1002/9781119211945.ch3

Costill, D. L., Cote, R., & Fink, W. (1976). Muscle water and electrolytes following varied levels of dehydration in man. *Journal of Applied Physiology, 40*(1), 6–11. https://doi.org/10.1152/jappl.1976.40.1.6

Depner, T. A. (1991). *Prescribing Hemodialysis: A Guide to Urea Modeling*. Springer US. https://doi.org/10.1007/978-1-4613-1509-4

Farkash, E. A., Wilson, A. M., & Jentzen, J. M. (2020). Ultrastructural Evidence for Direct Renal Infection with SARS-CoV-2. *Journal of the American Society of Nephrology: JASN, 31*(8), 1683–1687. https://doi.org/10.1681/ASN.2020040432

Fisch, C. (1973). Relation of Electrolyte Disturbances to Cardiac Arrhythmias. *Circulation, 47*(2), 408–419. https://doi.org/10.1161/01. CIR.47.2.408

Geokas, M. C., Lakatta, E. G., Makinodan, T., & Timiras, P. S. (1990). The Aging Process. *Annals of Internal Medicine, 113*(6), 455–466. https://doi.org/10.7326/0003-4819-113-6-455

González-Cuadrado, S., Lorz, C., García del Moral, R., O'Valle, F., Alonso, C., Ramiro, F., Ortiz-González, A., Egido, J., & Ortiz, A. (1997). Agonistic anti-Fas antibodies induce glomerular cell apoptosis in mice in vivo. *Kidney International, 51*(6), 1739–1746. https://doi. org/10.1038/ki.1997.239

Himmelfarb, J., & Ikizler, T. A. (2010). Hemodialysis. *New England Journal of Medicine, 363*(19), 1833–1845. https://doi.org/10.1056/ NEJMra0902710

Houston, M. C. (2011). The Importance of Potassium in Managing Hypertension. *Current Hypertension Reports, 13*(4), 309–317. https:// doi.org/10.1007/s11906-011-0197-8

Imdad, A., Jabeen, A., & Bhutta, Z. A. (2011). Role of calcium supplementation during pregnancy in reducing risk of developing gestational hypertensive disorders: A meta-analysis of studies from developing countries. *BMC Public Health, 11*(Suppl 3), S18. https://doi. org/10.1186/1471-2458-11-S3-S18

Iqbal, S., Klammer, N., & Ekmekcioglu, C. (2019). The Effect of Electrolytes on Blood Pressure: A Brief Summary of Meta-Analyses. *Nutrients, 11*(6), 1362. https://doi.org/10.3390/nu11061362

Kaplan, L. J., & Kellum, J. A. (2010). Fluids, pH, ions and electrolytes. *Current Opinion in Critical Care, 16*(4), 323–331. https://doi. org/10.1097/MCC.0b013e32833c0957

Khanam, F., Hossain, B., Mistry, S. K., Mitra, D. K., Raza, W. A.,

Rifat, M., Afsana, K., & Rahman, M. (2018). The association between daily 500 mg calcium supplementation and lower pregnancy-induced hypertension risk in Bangladesh. *BMC Pregnancy and Childbirth, 18*. https://doi.org/10.1186/s12884-018-2046-0

Knochel J. P. Clinical expression of potassium disturbances. In: Seldin DW, Giebisch G, eds. *The Regulation of Potassium Balance.* New York: Raven Press; 1989: 207-240.

Lobo, D. N. (2004). Fluid, electrolytes and nutrition: Physiological and clinical aspects. *Proceedings of the Nutrition Society, 63*(3), 453–466. https://doi.org/10.1079/PNS2004376

Locatelli, F., Manzoni, C., & Di Filippo, S. (2002). The importance of convective transport. *Kidney International. Supplement, 80*, 115–120. https://doi.org/10.1046/j.1523-1755.61.s80.21.x

Luckey, A. E., & Parsa, C. J. (2003). Fluid and Electrolytes in the Aged. *Archives of Surgery, 138(*10), 1055–1060. https://doi.org/10.1001/archsurg.138.10.1055

Marx, G., Meybohm, P., Schuerholz, T., Lotz, G., Ledinko, M., Schindler, A. W., Rossaint, R., & Zacharowski, K. (2019). Impact of a new balanced gelatine on electrolytes and pH in the perioperative care. *PLOS ONE, 14*(4), e0213057. https://doi.org/10.1371/journal.pone.0213057

Maughan, R. J. (2003). Impact of mild dehydration on wellness and on exercise performance. *European Journal of Clinical Nutrition, 57*(2), S19–S23. https://doi.org/10.1038/sj.ejcn.1601897

Nahum, J., Morichau-Beauchant, T., Daviaud, F., Echegut, P., Fichet, J., Maillet, J.-M., & Thierry, S. (2020). Venous Thrombosis Among Critically Ill Patients With Coronavirus Disease 2019 (COVID-19). *JAMA Network Open, 3*(5), e2010478–e2010478. https://doi.org/10.1001/jamanetworkopen.2020.10478

Nanovic, L. (2005). Electrolytes and Fluid Management in Hemodialysis and Peritoneal Dialysis. *Nutrition in Clinical Practice, 20*(2), 192–201. https://doi.org/10.1177/0115426505020002192

Pahl, M. V., Vaziri, N. D., Akbarpour, F., Afrasiabi, A., & Friis, R.

(1988). Effect of Rapid Weight Loss With Supplemented Fasting on Serum Electrolytes, Lipids, and Blood Pressure. *Journal of the National Medical Association, 80*(7), 803–809.

Pavenstädt, H., Kriz, W., & Kretzler, M. (2003). Cell Biology of the Glomerular Podocyte. *Physiological Reviews, 83*(1), 253–307. https://doi.org/10.1152/physrev.00020.2002

Powell, C. R., Stoller, M. L., Schwartz, B. F., Kane, C., Gentle, D. L., Bruce, J. E., & Leslie, S. W. (2000). Impact of body weight on urinary electrolytes in urinary stone formers22The views expressed in this article are those of the author and do not reflect the official policy or position of the United States Army, Department of Defense, or the U.S. Government. *Urology, 55*(6), 825–830. https://doi.org/10.1016/S0090-4295(99)00617-2

Rinschen, M. M., Benzing, T., Limbutara, K., & Pisitkun, T. (2015). Proteomic analysis of the kidney filtration barrier—Problems and perspectives. PROTEOMICS – *Clinical Applications, 9*(11–12), 1053–1068. https://doi.org/10.1002/prca.201400201

Sanz, A. B., Sanchez-Niño, M. D., & Ortiz, A. (2011). TWEAK, a multifunctional cytokine in kidney injury. *Kidney International, 80*(7), 708–718. https://doi.org/10.1038/ki.2011.180

Stevens, L. A., Coresh, J., Greene, T., & Levey, A. S. (2006). Assessing Kidney Function—Measured and Estimated Glomerular Filtration Rate. *New England Journal of Medicine, 354*(23), 2473–2483. https://doi.org/10.1056/NEJMra054415

Surawicz, B. (1966). Role of electrolytes in etiology and management of cardiac arrhythmias. *Progress in Cardiovascular Diseases, 8*(4), 364–386. https://doi.org/10.1016/S0033-0620(66)80011-7

Watson R L, Langford H G, Abernethy J, Barnes T Y, & Watson M J. (1980). Urinary electrolytes, body weight, and blood pressure. Pooled cross-sectional results among four groups of adolescent females. *Hypertension, 2*(4_pt_2), 193-98. https://doi.org/10.1161/01.HYP.2.4_Pt_2.193

Zamora, M. L., Tracy, B. L., Zielinski, J. M., Meyerhof, D. P., & Moss, M. A. (1998). Chronic Ingestion of Uranium in Drinking

Water: A Study of Kidney Bioeffects in Humans. *Toxicological Sciences, 43*(1), 68–77. https://doi.org/10.1093/toxsci/43.1.68

Chapter 3.

Catharine Ross, A., Ziegler, T. R., Tucker, K. L., Cousins, R. J., & Caballero, B. H. (2012). *Modern Nutrition in Health and Disease: Eleventh Edition.* Wolters Kluwer Health Adis.

Chawla, A., Sterns, R. H., Nigwekar, S. U., & Cappuccio, J. D. (2011). Mortality and Serum Sodium: Do Patients Die from or with Hyponatremia? *Clinical Journal of the American Society of Nephrology, 6(5)*, 960–965. https://doi.org/10.2215/cjn.10101110

Fong, J., & Khan, A. (2012). Hypocalcemia: updates in diagnosis and management for primary care. *Canadian family physician Medecin de famille canadien, 58*(2), 158–162.

Kardalas, E., Paschou, S. A., Anagnostis, P., Muscogiuri, G., Siasos, G., & Vryonidou, A. (2018). Hypokalemia: a clinical update. *Endocrine Connections, 7*(4). https://doi.org/10.1530/ec-18-0109

Lee, J. W. (2010). Fluid and Electrolyte Disturbances in Critically Ill Patients. *Electrolytes & Blood Pressure, 8*(2), 72. https://doi.org/10.5049/ebp.2010.8.2.72

Lodish, H., Berk, A., & Zipursky, S. L. (2000). Section 21.2 The Action Potential and Conduction of Electric Impulses. In *Molecular Cell Biology* (4th ed.). W. H. Freeman.

Stoppler, M. C. (2019, December 24). *What Do Electrolytes Do? Benefits, Chemistry & Imbalance Symptoms.* MedicineNet. https://www.medicinenet.com/electrolytes/article.htm.

Sweeney, H. L., & Hammers, D. W. (2018). Muscle Contraction. *Cold Spring Harbor Perspectives in Biology, 10*(2). https://doi.org/10.1101/cshperspect.a023200

Upadhyay, A., Jaber, B. L., & Madias, N. E. (2006). Incidence and Prevalence of Hyponatremia. *The American Journal of Medicine, 119*(7). https://doi.org/10.1016/j.amjmed.2006.05.005

Weiss-Guillet, E.-M., Takala, J., & Jakob, S. M. (2003). Diagnosis and management of electrolyte emergencies. *Best Practice & Research Clinical Endocrinology & Metabolism, 17*(4), 623–651. https://doi.org/10.1016/s1521-690x(03)00056-3

Chapter 4.

Anatomy and Physiology. (2019). OSCRiceUniversity. https://opentextbc.ca/anatomyandphysiologyopenstax/.

Berry, J. (2020). *Electrolyte imbalance: Symptoms, causes and treatment.* Medical News Today. https://www.medicalnewstoday.com/articles/electrolyte-imbalance#in-older-adults.

Buchanan, K. (2020, November 9). *WHAT HAPPENS WHEN YOUR BODY IS LOW ON ELECTROLYTES?* GOODONYA® Organic. https://goodonyaorganic.com/blogs/goodonya-hydrate/what-happens-when-your-body-is-low-on-electrolytes.

Canada's Food Guide. (2021). Government of Canada. https://food-guide.canada.ca/en/

Dr. Dave (2018). What Are Electrolytes? YouTube. https://www.youtube.com/watch?v=a6Dw9vtnwns&t=11s&ab_channel=ProfessorDaveExplains.

El-Sharkawy, A. M., Sahota, O., Maughan, R. J., & Lobo, D. N. (2013, November 22). *The pathophysiology of fluid and electrolyte balance in the older adult surgical patient.* Clinical Nutrition. https://www.sciencedirect.com/science/article/pii/S0261561413003166.

Holland, K. (2019, April 29). *What You Need to Know About Electrolyte Disorders.* Healthline. https://www.healthline.com/health/electrolyte-disorders.

MediLexicon International. (n.d.). *Electrolyte imbalance: Symptoms, causes and treatment.* Medical News Today. https://www.medicalnewstoday.com/articles/electrolyte-imbalance#treatment.

Moore, M (2021). *Do Carbohydrates Cause Weight Gain.* Sharecare. https://www.sharecare.com/health/carbohydrate-body-impact/do-carbohydrates-cause-weight-gain.

Oxford Languages. (2021). Oxford English Dictionary. https://languages.oup.com/research/oxford-english-dictionary/.

Shrimanker, I. (2020, September 12). *Electrolytes*. StatPearls. https://www.ncbi.nlm.nih.gov/books/NBK541123/.

Templeton, H. (2020, October 26). *Skip Sugary Sports Drink and Replenish Your Electrolytes With These Foods*. Runner's World.https://www.runnersworld.com/nutrition-weight-loss/a20795183/the-best-foods-to-replenish-electrolytes/.

U.S. National Library of Medicine. (2021, April 23). *Fluid and Electrolyte Balance*. MedlinePlus. https://medlineplus.gov/fluidandelectrolytebalance.html.

What are Electrolytes? Cedars. (n.d.). https://www.cedars-sinai.org/blog/electrolytes.html.

Whelan, C. (2019, May 13). *Electrolytes Food: 25 Foods for Electrolyte Function and More*. Healthline. https://www.healthline.com/health/fitness-nutrition/electrolytes-food#what-are-electrolytes.

Chapter 5.

American College of Sports Medicine, Sawka, M. N., Burke, L. M., Eichner, E. R., Maughan, R. J., Montain, S. J., & Stachenfeld, N. S. (2007). American College of Sports Medicine position stand. Exercise and fluid replacement. *Medicine and science in sports and exercise, 39*(2), 377–390. https://doi.org/10.1249/mss.0b013e31802ca597

Armstrong, L. E. (1994). Considerations for Replacement Beverages: Fluid-Electrolyte Balance and Heat Illness. *Fluid Replacement and Heat Stress*. https://doi.org/10.17226/9071

Baker L. B. (2017). Sweating Rate and Sweat Sodium Concentration in Athletes: A Review of Methodology and Intra/Interindividual Variability. *Sports medicine (Auckland, N.Z.), 47*(Suppl 1), 111–128. https://doi.org/10.1007/s40279-017-0691-5

Field, A. E., Sonneville, K. R., Falbe, J., Flint, A., Haines, J., Rosner, B., & Camargo, C. A. (2014). Association of sports drinks with weight gain among adolescents and young adults. *Obesity, 22*(10), 2238–2243.

https://doi.org/10.1002/oby.20845

Food Data Central. (n.d.). https://fdc.nal.usda.gov/.

Glazer J. L. (2005). Management of heatstroke and heat exhaustion. *American family physician, 71*(11), 2133–2140.

Grillo, A., Salvi, L., Coruzzi, P., Salvi, P., & Parati, G. (2019). Sodium Intake and Hypertension. *Nutrients, 11*(9), 1970. https://doi. org/10.3390/nu11091970

Hayes, P. A., Fraher, M. H., & Quigley, E. M. (2014). Irritable bowel syndrome: the role of food in pathogenesis and management. *Gastroenterology & hepatology, 10*(3), 164–174.

Iqbal, S., Klammer, N., & Ekmekcioglu, C. (2019). The Effect of Electrolytes on Blood Pressure: A Brief Summary of Meta-Analyses. *Nutrients, 11*(6), 1362. https://doi.org/10.3390/nu11061362

Johnson, R. K., Appel, L. J., Brands, M., Howard, B. V., Lefevre, M., Lustig, R. H., Sacks, F., Steffen, L. M., Wylie-Rosett, J., & American Heart Association Nutrition Committee of the Council on Nutrition, Physical Activity, and Metabolism and the Council on Epidemiology and Prevention (2009). *Dietary sugars intake and cardiovascular health: a scientific statement from the American Heart Association, 120*(11), 1011–1020. https://doi.org/10.1161/CIRCULATIONAHA.109.192627

Lau, W. Y., Kato, H., & Nosaka, K. (2021). Effect of oral rehydration solution versus springwater intake during exercise in the heat on muscle cramp susceptibility of young men. *Journal of the International Society of Sports Nutrition, 18*(1). https://doi.org/10.1186/s12970-021-00414-8

Lipman, G. S., Burns, P., Phillips, C., Jensen, J., Little, C., Jurkiewicz, C., ... Krabak, B. J. (2020). Effect of Sodium Supplements and Climate on Dysnatremia During Ultramarathon Running. *Clinical Journal of Sport Medicine, Publish Ahead of Print.* https://doi. org/10.1097/jsm.0000000000000832

Nazarian, L. F. (1997). A Synopsis of the American Academy of Pediatrics' Practice Parameter 221–223. https://doi.org/10.1542/pir.18-

Orrù, S., Imperlini, E., Nigro, E., Alfieri, A., Cevenini, A., Polito, R., Daniele, A., Buono, P., & Mancini, A. (2018). Role of Functional Beverages on Sport Performance and Recovery. *Nutrients, 10*(10), 1470. https://doi.org/10.3390/nu10101470

Park, S., Blanck, H. M., Sherry, B., Brener, N., & O'Toole, T. (2012). Factors associated with sugar-sweetened beverage intake among United States high school students. *The Journal of nutrition, 142*(2), 306–312. https://doi.org/10.3945/jn.111.148536

Pedialyte. (n.d.). *Pedialyte® Fruit.* Pedialyte® with zinc - Fruit Flavour. https://pedialyte.ca/en/products/liters/fruit#:~:text=Serving%20 size%20%3A%201%20L %20(1000%20mL)&text=MEDICINAL%20 INGREDIENTS%3A%20Dextrose%2C%2c itric%20acid,6.

Pound, C. M., & Blair, B. (2018). Re: Pound CM, Blair B. Energy and sports drinks in children and adolescents. Paediatr Child Health. 2017;22(7):406–10. *Paediatrics & Child Health, 23*(4), 302–302. https://doi.org/10.1093/pch/pxy065

Prescribers' Digital Reference. (n.d.). Essentia Water | FULL Prescribing Information. *Essentia Water | FULL Prescribing Information | PDR.net.* https://www.pdr.net/full-prescribing-information/essentia-water?druglabelid=3618.

RxList. (2021, April 29). *Pedialyte: Uses, Side Effects, Interactions & Pill Images.* RxList. https://www.rxlist.com/fdb/drugs/2901/pedialyte-brand.htm.

Spencer, M., Gupta, A., Dam, L. V., Shannon, C., Menees, S., & Chey, W. D. (2016). Artificial Sweeteners: A Systematic Review and Primer for Gastroenterologists. *Journal of neurogastroenterology and motility, 22*(2), 168–180. https://doi.org/10.5056/jnm15206

Chapter 6.

Allison, S. (2004). Fluid, electrolytes and nutrition. *Clinical Medicine, 4*(6), 573–578. https://doi.org/10.7861/clinmedicine.4-6-573

Beswick, K. (2019). *What are Electrolytes?* Cedars-Sinai. https://www.cedars-sinai.org/blog/electrolytes.html

Born, S. (n.d.). *Electrolyte replenishment: Why it's so important and how to do it right.* Hammer Nutrition. Retrieved May 8, 2021, from https://www.hammernutrition.com/knowledge/advanced-knowledge/ electrolyte-replenishment-why-its-important-and-how-to-do-it-right

Brouns, F., & Kovacs, E. (1997). Functional drinks for athletes. *Trends in Food Science & Technology, 8*(12), 414–421. https://doi.org/10.1016/ s0924-2244(97)01098-4

Convertino, V. A. (1987). Fluid shifts and hydration state: effects of long-term exercise. *Can J Sport Sci, 12*, 136S-139S.

Felman, A. (2017, November 20). *Everything you need to know about electrolytes.* Medical News Today. https://www.medicalnewstoday.com/ articles/153188

Hinchcliff, K. W., Geor, R. J., & Kaneps, A. J. (2008). *Equine Exercise Physiology: The Science of Exercise in the Athletic Horse.* Elsevier.

Mineral Resources International. (2007). *Electrolytes and human health.* http://www.mineralresourcesint.co.uk/pdf/products/Electrolyte_line_ HS.pdf

Orrù, S., Imperlini, E., Nigro, E., Alfieri, A., Cevenini, A., Polito, R., Daniele, A., Buono, P., & Mancini, A. (2018). Role of Functional Beverages on Sport Performance and Recovery. *Nutrients, 10*(10), 1470. https://doi.org/10.3390/nu10101470

Royal Society of Chemistry. (2020). *The chemistry of sports drinks: Teacher's notes.* https://edu.rsc.org/download?ac=12462

Shirreffs, S. M. (2009). Hydration in sport and exercise: water, sports drinks and other drinks. *Nutrition Bulletin, 34*(4), 374–379. https://doi. org/10.1111/j.1467-3010.2009.01790.x

The science behind electrolytes. (n.d.). Elete Electrolyte. Retrieved May 8, 2021, from https://eletewater.co.uk/pages/the-science-behind-electrolytes#.YJbmIbVKjIU

Chapter 7.

Bhargava, H. D. (2020). *Hyponatremia: Symptoms, Causes, and Treatments.* WebMD. https://www.webmd.com/a-to-z-guides/what-is-hyponatremia#1-4.

Busch, S. (2018, December 14). *What Are the Benefits of Electrolyte Water?* SF Gate. https://healthyeating.sfgate.com/benefits-electrolyte-water-12128.html.

Cleveland Clinic. (2020, September 18). *Electrolyte Drinks: Beneficial or Not?* Health Essentials from Cleveland Clinic. https://health.clevelandclinic.org/electrolyte-drinks-beneficial-or-not/.

Coombes, J. S., & Hamilton, K. L. (2000). The Effectiveness of Commercially Available Sports Drinks. *Sports Medicine, 29*(3), 181–209. https://doi.org/10.2165/00007256-200029030-00004

Coso, J. D., Estevez, E., Baquero, R. A., & Mora-Rodriguez, R. (2008). Anaerobic performance when rehydrating with water or commercially available sports drinks during prolonged exercise in the heat. *Applied Physiology, Nutrition, and Metabolism, 33*(2), 290–298. https://doi.org/10.1139/h07-188

Davis, J. M., Lamb, D. R., Pate, R. R., Slentz, C. A., Burgess, W. A., & Bartoli, W. P. (1988). Carbohydrate-electrolyte drinks: effects on endurance cycling in the heat. *The American Journal of Clinical Nutrition, 48*(4), 1023–1030. https://doi.org/10.1093/ajcn/48.4.1023

Dawes, J. J., Campbell, B. I., Ocker, L. B., Temple, D. R., Carter, J. G., & Brooks, K. A. (2014). The Effects of a Pre-Workout Energy Drink on Measures of Physical Performance. *International Journal of Physical Education, Fitness and Sports, 3*(4), 122–131. https://doi.org/10.26524/14420

Desbrow, B., Jansen, S., Barrett, A., Leveritt, M. D., & Irwin, C. (2014). Comparing the rehydration potential of different milk-based drinks to a carbohydrate–electrolyte beverage. *Applied Physiology, Nutrition, and Metabolism, 39*(12), 1366–1372. https://doi.org/10.1139/apnm-2014-0174

Dragos-Florin, T. (2017). Hydration in tennis performance – water, carbohydrate or electrolyte sports drink? *Science, Movement and Health,* *17*(2), 511–516.

DiSilvestro, R. A. (2011). Effects of a Mixed Nutraceutical Beverage on Performance of Moderately Strenuous Aerobic Exercise Lasting Under an Hour. *The Open Nutraceuticals Journal, 4*(1), 151–155. https://doi.org/10.2174/1876396001104010151

Hornsby, J. (2011). The effects of carbohydrate-electrolyte sports drinks on performance and physiological function during an 8km cycle time trial. *The Plymouth Student Scientist, 4*(2), 30–49.

Jeukendrup, A., Brouns, F., Wagenmakers, A., & Saris, W. (1997). Carbohydrate-Electrolyte Feedings Improve 1 h Time Trial Cycling Performance. *International Journal of Sports Medicine, 18*(02), 125–129. https://doi.org/10.1055/s-2007-972607

Kalman, D. S., Feldman, S., Krieger, D. R., & Bloomer, R. J. (2012). Comparison of coconut water and a carbohydrate-electrolyte sport drink on measures of hydration and physical performance in exercise-trained men. *Journal of the International Society of Sports Nutrition, 18*(2). https://doi.org/https://doi.org/10.1186/1550-2783-9-1

Leiper, J. (1998). Intestinal Water Absorption - Implications for the Formulation of Rehydration Solutions. *International Journal of Sports Medicine, 19*(S 2). https://doi.org/10.1055/s-2007-971977

Manna, I., & Khanna, G. L. (2005). Supplementary Effect of Creatine on Cardiovascular Adaptation and Endurance Performance in Athletes. *Indian J Med Res, 1*(1), 665–669. https://doi.org/10.4172/2473-6449.1000106

Maughan, R. J., Leiper, J. B., & Shirreffs, S. M. (1997). Factors influencing the restoration of fluid and electrolyte balance after exercise in the heat. *British Journal of Sports Medicine, 31*(3), 175–182. https://doi.org/10.1136/bjsm.31.3.175

Maughan, R. J., Owen, J. H., Shirreffs, S. M., & Leiper, J. B. (1994). Post-exercise rehydration in man: effects of electrolyte addition to ingested fluids. *European Journal of Applied Physiology and Occupational Physiology, 69*(3), 209–215. https://doi.org/10.1007/bf01094790

McRae, K. A., & Galloway, S. D. R. (2012). Carbohydrate-Electrolyte Drink Ingestion and Skill Performance During and After 2 hr of Indoor Tennis Match Play. *International Journal of Sport Nutrition and Exercise Metabolism, 22*(1), 38–46. https://doi.org/10.1123/ijsnem.22.1.38

Meckes, N., & Brown, S. (2017). The effects of carbohydrate-electrolyte drinks on physical and mental performance. *Journal of Science and Medicine in Sport, 20*, 23–25. https://doi.org/10.1016/j.jsams.2016.12.028

Meixner, M. (2018). *Electrolyte Water: Benefits and Myths*. Healthline. https://www.healthline.com/nutrition/electrolyte-water.

Miller, K. C., Mack, G., & Knight, K. L. (2009). Electrolyte and Plasma Changes After Ingestion of Pickle Juice, Water, and a Common Carbohydrate-Electrolyte Solution. *Journal of Athletic Training, 44*(5), 454–461. https://doi.org/10.4085/1062-6050-44.5.454

Nassis, G. P., Williams, C., & Chisnall, P. (1998). Effect of a carbohydrate-electrolyte drink on endurance capacity during prolonged intermittent high intensity running. *Br J Sports Med, 32*, 248–252. https://doi.org/10.1136/bjsm/32.3.248

Nielsen, B., SjØgaard, G., Ugelvig, J., Knudsen, B., & Dohlmann, B. (1986). Fluid balance in exercise dehydration and rehydration with different glucose-electrolyte drinks. *European Journal of Applied Physiology and Occupational Physiology, 55*(3), 318–325. https://doi.org/10.1007/bf02343806

Ostojic, S. M., & Mazic, S. (2002). Effects of a Carbohydrate-Electrolyte Drink on Specific Soccer Tests and Performance. *J Sports Sci Med, 1*(2), 47–53.

Pérez-Idárraga, A., & Aragón-Vargas, L. F. (2014). Postexercise rehydration: potassium-rich drinks versus water and a sports drink. *Applied Physiology, Nutrition, and Metabolism, 39*(10), 1167–1174. https://doi.org/10.1139/apnm-2013-0434

Powers, S. K., Lawler, J., Dodd, S., Tulley, R., Landry, G., & Wheeler, K. (1990). Fluid replacement drinks during high intensity exercise effects on minimizing exercise-induced disturbances in homeostasis.

European Journal of Applied Physiology and Occupational Physiology,
60(1), 54–60. https://doi.org/10.1007/bf00572186

Raizel, R., Coqueiro, A. Y., Bonvini, A., & Tirapegui, J. (2019). Sports
and Energy Drinks: Aspects to Consider. In A. M. Grumezescu &
A. M. Holban (Eds.), *The science of beverages* (Vol. 10, pp. 1–30). essay,
Woodhead Publishing.

Roberts, A. (2017). The addition of electrolytes to a carbohydrate-
based sport drink: Effect on continuous incremental exercise done
against progressively greater workloads. Kentucky Association of
Health, Physical Education, Recreation and Dance. 55. 39-48.

Rosner, M. H., & Bennett, B. (2007). Exercise-Associated
Hyponatremia. *Clinical Journal of the American Society of Nephrology, 2,*
151–161.doi: 10.2215/CJN.02730806

Sampaio de Melo, M. A., Passos, V. F., Lima, J. P., Santiago, S. L., &
Rodrigues, L. K. (2016). Carbohydrate-electrolyte drinks exhibit risks
for human enamel surface loss. *Restorative Dentistry & Endodontics,*
41(4), 246–254. https://doi.org/10.5395/rde.2016.41.4.246

Shephard, R. H. (2019). Sports and Energy Drinks: Aspects to
Consider. In A. M. Grumezescu & A. M. Holban (Eds.), *The science of*
beverages (Vol. 10, pp. 1–30). essay, Woodhead Publishing.

Shirreffs, S. M., Armstrong, L. E., & Cheuvront, S. N. (2007). Fluid
and electrolyte needs for preparation and recovery from training
and competition. *Journal of Sports Sciences, 22*(1), 57–63. https://doi.
org/10.4324/9780203448618_chapter_5

Shirreffs, S. M., & Sawka, M. N. (2013). Fluid and electrolyte needs
for training, competition, and recovery. *Journal of Sports Sciences, 29,*
47–54. https://doi.org/10.4324/9781315873268-12

Singh, A., Sarika, C., & Sandhu, J. S., (2011). Efficacy of pre exercise
carbohydrate drink (Gatorade) on the recovery heart rate, blood lactate
and glucose levels in short term intensive exercise. ,Serbian Journal of
Sports Sciences. 5. 29-34.

Tharion, W. J., Montain, S. J., O'Brien, C., Shippee, R. L., & Hoban,
J. L. (1997). Effects of military exercise tasks and a carbohydrate-

electrolyte drink on rifle shooting performance in two shooting positions. *International Journal of Industrial Ergonomics, 19*(1), 31–39. https://doi.org/10.1016/0169-8141(95)00087-9

Tinsley, G. (2018). *Sports Drinks: Should You Drink Them Instead of Water?* Healthline. https://www.healthline.com/nutrition/sports-drinks.

Tsintzas, O. K., Williams, C., Singh, R., Wilson, W., & Burrin, J. (1995). Influence of carbohydrate-electrolyte drinks on marathon running performance. *European Journal of Applied Physiology and Occupational Physiology, 70*(2), 154–160. https://doi.org/10.1007/bf00361543

Watson, P., Love, T. D., Maughan, R. J., & Shirreffs, S. M. (2008). A comparison of the effects of milk and a carbohydrate-electrolyte drink on the restoration of fluid balance and exercise capacity in a hot, humid environment. *European Journal of Applied Physiology, 104*(4), 633–642. https://doi.org/10.1007/s00421-008-0809-4

Wortley, G., & Islas, A. A. (2011). The problem with ultra-endurance athletes. *British Journal of Sports Medicine, 45*(14), 1085–1085. https://doi.org/10.1136/bjsports-2011-090399

Chapter 8.

Baker, L. B. (2017). Sweating rate and sweat sodium concentration in athletes: A review of methodology and Intra/Interindividual Variability. *Sports Medicine, 47*(S1), 111-128. doi:10.1007/s40279-017-0691-5

Baker, L. B., De Chavez, P. J., Ungaro, C. T., Sopeña, B. C., Nuccio, R. P., Reimel, A. J., & Barnes, K. A. (2018). Exercise intensity effects on total sweat electrolyte losses and regional vs. whole-body sweat [na+], [cl–], and [k+]. *European Journal of Applied Physiology, 119*(2), 361-375. doi:10.1007/s00421-018-4048-z

Banwell, J. G., Pierce, N. F., Mitra, R. C., Brigham, K. L., Caranasos, G. J., Keimowitz, R. I., . . . Mondal, A. (1970). Intestinal fluid and electrolyte transport in human cholera. *Journal of Clinical Investigation, 49*(1), 183-195. doi:10.1172/jci106217

Betts, J., Young, K., Wise, J., Johnson, E., Poe, B., Kruse, D., . . . DeSaix, P. (2013, March 06). *Electrolyte Balance*. Retrieved May 04, 2021, from https://opentextbc.ca/anatomyandphysiologyopenstax/chapter/electrolyte-balance/

Collins, J., Nguyen, A., & Badireddy, M. (2020, August 10). Anatomy, abdomen and pelvis, small intestine. Retrieved May 06, 2021, from https://www.ncbi.nlm.nih.gov/books/NBK459366/

Darwish, A., & Lui, F. (2020, October 03). Physiology, colloid osmotic pressure. Retrieved May 06, 2021, from https://www.ncbi.nlm.nih.gov/books/NBK541067/

Edelman, I., & Leibman, J. (1959). Anatomy of body water and electrolytes. *The American Journal of Medicine, 27*(2), 256-277. doi:10.1016/0002-9343(59)90346-8

Field, M. (2003). Intestinal ion transport and the pathophysiology of diarrhea. *Journal of Clinical Investigation, 111*(7), 931-943. doi:10.1172/jci200318326

Gerritsen, K., Boer, W., & Joles, J. (2015). The importance of intake: a gut feeling. *Annals Of Translational Medicine, 3*(4), 5. doi:10.3978/j.issn.2305-5839.2015.03.21

Hodge, B., Sanvictores, T., & Brodell, R. T. (2020, October 01). Anatomy, skin sweat glands. Retrieved May 04, 2021, from https://www.ncbi.nlm.nih.gov/books/NBK482278/

Hooton, D., Lentle, R., Monro, J., Wickham, M., & Simpson, R. (2015). The secretion and action of Brush Border enzymes in the Mammalian small intestine. Reviews of Physiology, Biochemistry and Pharmacology, 59-118. doi:10.1007/112_2015_24

Kiela, P. R., & Ghishan, F. K. (2016). Physiology of intestinal absorption and secretion. *Best Practice & Research Clinical Gastroenterology, 30*(2), 145-159. doi:10.1016/j.bpg.2016.02.007

OpenStax. (2013, April 12). Anatomy and physiology ii. Retrieved May 05, 2021, from https://courses.lumenlearning.com/suny-ulster-ap2/chapter/electrolyte-balance/

OpenStax. (2018, March 31). Anatomy and physiology ii. Retrieved May 05, 2021, from https://courses.lumenlearning.com/suny-ap2/chapter/body-fluids-and-fluid-compartments-no-content/

Sakuyama, H., Katoh, M., Wakabayashi, H., Zulli, A., Kruzliak, P., & Uehara, Y. (2016). Influence of gestational salt restriction in fetal growth and in development of diseases in adulthood. *Journal of Biomedical Science, 23*(1). doi:10.1186/s12929-016-0233-8

Scott, J., Menouar, M. A., & Dunn, R. J. (2021, February 15). Physiology, aldosterone. Retrieved May 03, 2021, from https://www.ncbi.nlm.nih.gov/books/NBK470339/

Wang, C., Cogswell, M. E., Loria, C. M., Chen, T., Pfeiffer, C. M., Swanson, C. A., . . . Burt, V. L. (2013). Urinary excretion of sodium, potassium, and chloride, but not iodine, varies by timing of collection in a 24-Hour calibration study. *The Journal of Nutrition, 143*(8), 1276-1282. doi:10.3945/jn.113.175927

Wang, F., Butler, T., Rabbani, G., & Jones, P. K. (1986). The acidosis of cholera. *New England Journal of Medicine, 315*(25), 1591-1595. doi:10.1056/nejm198612183152506

Yousef, H., Alhajj, M., & Sharma, S. (2020, July 27). Anatomy, skin (Integument), Epidermis. Retrieved May 06, 2021, from https://www.ncbi.nlm.nih.gov/books/NBK470464/

Chapter 9.

BBB National Programs. (2020, September 30). *NAD Finds Certain BodyArmor SuperDrink Claims Supported; Recommends Discontinuance of "The Only Sports Drink" Claims in the Challenged Ads.* https://bbbprograms.org/media-center/newsroom/nad-finds-certain-bodyarmor-superdrink-claims-supported-recommends-discontinuance-of-the-only-sports-drink-claims-in-the-challenged-ads

BBC News. (2014, January 8). *Lucozade Sport drink advert banned after complaints.* https://www.bbc.com/news/newsbeat-25650519

Brison, N. T., Baker, T. A., Byon, K. K., & Evans, N. J. (2020). An Interdisciplinary Examination of the Material Effects of Deceptive Sport Beverage Advertisements. *Journal of Global Sport Management ;*

Page 1-22 ; ISSN 2470-4067 2470-4075. https://doi.org/10.1080/24704
067.2020.1711531

Cohen, D. (2014). Sports drinks adverts are banned for false claims.
BMJ: British Medical Journal, 348. https://doi.org/10.1136/bmj.g136

Cohen, D. (2012). The truth about sports drinks. *BMJ: British Medical
Journal, 345(7866),* 20–25. https://doi.org/10.1136/bmj.e4737

Consumer Law Group. (n.d.). *Vita Coco coconut water National Class
Action.* https://www.clg.org/Class-Action/List-of-Class-Actions/Vita-
Coco-coconut-water-National-Class-Action#form

Conway, J. (2021, April 22). *Sports drink unit sales in U.S. convenience
stores (C-stores) in 2020, by brand.* Statista. https://www.statista.com/
statistics/408976/best-selling-sports-drink-brands-in-us-c-stores-
based-on-unit-sales/

Dreyfuss, J. H. (2016, October 4). *Popular Commercial Sports Drinks
Are a Scam.* MDalert.com. https://www.mdalert.com/article/popular-
commercial-sports-drinks-are-a-scam

Earley, M. (2019, June 3). *Powerade advert featuring Israel Dagg and
Steven Adams taken off air over misleading claims.* Stuff. https://www.
stuff.co.nz/entertainment/113196662/powerade-advert-featuring-israel-
dagg-and-steven-adams-taken-off-air-over-misleading-claims

Edelstein, J. (2018, December 11). *NAD Refers BodyArmor Claims to
FTC After Review.* Manatt, Phelps & Phillips, LLP. https://www.
jdsupra.com/legalnews/nad-refers-bodyarmor-claims-to-ftc-67938/

Fortune Business Insights. (2021, February 24). *Sports Drink Market
Size to Hit $32.61 Billion by 2026 | Fortune Business Insights™.* https://
www.globenewswire.com/news-release/2021/02/24/2181211/0/en/
Sports-Drink-Market-Size-to-Hit-32-61-Billion-by-2026-Fortune-
Business-Insights.html

Miller, K. C. (2014). Electrolyte and plasma responses after pickle
juice, mustard, and deionized water ingestion in dehydrated
humans. *Journal of athletic training, 49(3),* 360–367. https://doi.
org/10.4085/1062-6050-49.2.23

Miller, K. C., Mack, G. W., Knight, K. L., Hopkins, J. T., Draper, D. O., Fields, P. J., & Hunter, I. (2010). Reflex inhibition of electrically induced muscle cramps in hypohydrated humans. *Medicine and science in sports and exercise, 42(5)*, 953–961. https://doi.org/10.1249/MSS.0b013e3181c0647e

Pound, C. M., & Blair, B. (2017). Energy and sports drinks in children and adolescents. *Paediatrics & Child Health (1205-7088), 22*(7), 406–410. https://doi.org/10.1093/pch/pxx132

Rothstein, R.Y. (2012, March 19). *Maker of Coconut Water Settles False Advertising Class Action.* Winston & Strawn, LLC. https://www.winston.com/en/advertising-marketing-privacy-law-news/maker-of-coconut-water-settles-false-advertising-class-action.html

Shaikh, T. (2014, January 28). *Lucozade Sport ad campaign banned for claiming drink hydrates better than water.* Independent. https://www.independent.co.uk/news/media/advertising/lucozade-sport-ad-campaign-banned-claiming-drink-hydrates-better-water-9047437.html

Shang, G., Collins, M., & Schwellnus, M. P. (2011). Factors associated with a self-reported history of exercise-associated muscle cramps in Ironman triathletes: a case-control study. *Clinical journal of sport medicine : official journal of the Canadian Academy of Sport Medicine, 21(3)*, 204–210. https://doi.org/10.1097/JSM.0b013e31820bcbfd

Schwellnus M. P. (2009). Cause of exercise associated muscle cramps (EAMC)--altered neuromuscular control, dehydration or electrolyte depletion?. *British journal of sports medicine, 43(6)*, 401–408. https://doi.org/10.1136/bjsm.2008.050401

Schwellnus, M. P., Drew, N., & Collins, M. (2011). Increased running speed and previous cramps rather than dehydration or serum sodium changes predict exercise-associated muscle cramping: a prospective cohort study in 210 Ironman triathletes. *British journal of sports medicine, 45(8)*, 650–656. https://doi.org/10.1136/bjsm.2010.078535

Theeboom, S. (2014, April 28). *The Top-Selling Coconut Water Brands in One Infographic.* First We Feast. https://firstwefeast.com/eat/2014/04/top-selling-coconut-water-brands-one-infographic

The New Zealand Herald. (2019, May 31). *Coca Cola Oceania forced to* remove 'misleading' Powerade ION4 ads featuring Kiwi sports stars. https://www.nzherald.co.nz/business/coca-cola-oceania-forced-to-remove-misleading-powerade-ion4-ads-featuring-kiwi-sports-stars/

The Pickle Juice Company. (n.d.-b). *Home.* https://picklepower.com/

The Pickle Juice Company. (n.d.-a). *Learn: The Science.* https://picklepower.com/pages/the-science

Thompson, M., Heneghan, C., & Cohen, D. (2012). How valid is the European Food Safety Authority's assessment of sports drinks? *BMJ: British Medical Journal, 345,* e4753. https://doi.org/10.1136/bmj.e4753

Tinsley, G. (2018, May 13). *Should You Drink Sports Drinks Instead of Water?* Healthline. https://www.healthline.com/nutrition/sports-drinks

Tuller, D. (2012, July 30). *Do Sports Drinks Really Work?.* Mother Jones. https://www.motherjones.com/politics/2012/07/do-sports-drinks-really-work/

Chapter 10.

Al-Shaar L, Vercammen K, Lu C, Richardson S, Tamez M, Mattei J. Health Effects and Public Health Concerns of Energy Drink Consumption in the United States: A Mini-Review. Front Public Health. 2017;5:225.

Baker, L. B., Rollo, I., Stein, K. W., & Jeukendrup, A. E. (2015). Acute Effects of Carbohydrate Supplementation on Intermittent Sports Performance. *Nutrients, 7*(7), 5733–5763. https://doi.org/10.3390/nu7075249

Broughton, D., Fairchild, R. M., & Morgan, M. Z. (2016). A survey of sports drinks consumption among adolescents. *British Dental Journal, 220*(12), 639–643. https://doi.org/10.1038/sj.bdj.2016.449

Brouns, F. J. P. H., Kovacs, E. M. R., & Senden, J. M. G. (1998). The effect of different rehydration drinks on post-exercise electrolyte excretion in trained athletes. International journal of sports medicine, 19(1), 56-60.

Callahan, P. (2014). What is the Difference Between Sport and Energy Drinks. *ThedaCare*. https://thedacare.org/news-and-events/healthy-living/what-is-the-difference-between-sport-and-energy-drinks/

Coombes, J. S., & Hamilton, K. L. (2000). The effectiveness of commercially available sports drinks. *Sports Medicine (Auckland, N.Z.), 29*(3), 181–209. https://doi.org/10.2165/00007256-200029030-00004

Drink Like A Champion—Celebrity Drink Endorsements. (n.d.). Retrieved May 8, 2021, from https://www.topendsports.com/nutrition/hydration/athlete-endorsements.htm

Fischer-Colbrie, M. (n.d.). *The Importance of Electrolytes for Athletes | BridgeAthletic.* Retrieved May 7, 2021, from https://blog.bridgeathletic.com/electrolytes-for-athletes

Healthy Eating Research. Consumption of sports drinks by children and adolescents. 2012. Available online at http://healthyeatingresearch.org/wp-content/uploads/2013/12/HER-Sports-Drinks-Research-Review-6-2012.pdf

Horrigan, M. (2016). Battle of the Brands: Gatorade vs. Powerade. *Soda Pop Culture.* https://sodapopculture.wordpress.com/2016/11/30/battle-of-the-brands-gatorade-vs-powerade/

Hyponatremia. (2015, December 24). National Kidney Foundation. https://www.kidney.org/atoz/content/hyponatremia

IOC. (n.d.). London 2012 Individual women Results—Olympic triathlon. Olympic Channel. Retrieved May 8, 2021, from https://www.olympicchannel.com/en/olympic-games/london-2012/triathlon/individual-women

Jeukendrup, A., Brouns, F., Wagenmakers, A. J., & Saris, W. H. (1997). Carbohydrate-electrolyte feedings improve 1 h time trial cycling performance. *International Journal of Sports Medicine, 18*(2), 125–129. https://doi.org/10.1055/s-2007-972607

Jeukendrup, A. E. (2011). Nutrition for endurance sports: Marathon, triathlon, and road cycling. *Journal of Sports Sciences, 29 Suppl 1,* S91-99. https://doi.org/10.1080/02640414.2011.610348

Kenefick, R. W., & Cheuvront, S. N. (2012). Hydration for recreational sport and physical activity. *Nutrition Reviews, 70 Suppl 2*, S137-142. https://doi.org/10.1111/j.1753-4887.2012.00523.x

ltd, M. D. F. (n.d.). *Electrolyte Drinks Market Growth, Size, Share and Forecast to 2025*. Market Data Forecast. Retrieved May 3, 2021, from http://www.marketdataforecast.com/

Orrù, S., Imperlini, E., Nigro, E., Alfieri, A., Cevenini, A., Polito, R., Daniele, A., Buono, P., & Mancini, A. (2018). Role of Functional Beverages on Sport Performance and Recovery. *Nutrients, 10*(10). https://doi.org/10.3390/nu10101470

NCCIH. (2018). Energy Drinks. https://www.nccih.nih.gov/health/energy-drinks

Nelson B. (2021, March 18.). *This is Why Some People Sweat More Than Others—And How to Manage It*. The Healthy. https://www.thehealthy.com/beauty/face-body-care/why-do-i-sweat-so-much/

Wootton-Beard, P. C., & Ryan, L. (2011). Improving public health?: The role of antioxidant-rich fruit and vegetable beverages. *Food Research International, 44*(10), 3135–3148. https://doi.org/10.1016/j.foodres.2011.09.015

Wyckoff, A.S. (2011). Sports drinks vs. energy drinks vs. plain water: What's best for thirsty kids? *APP News & Journals Gateway, 32*(6). https://doi.org/10.1542/aapnews.2011326-32a

Chapter 11.

Canavan, A., & Arant, B. S. (2009). Diagnosis and Management of Dehydration in Children. *American Family Physician, 80*(7), 692–696. https://www.aafp.org/afp/2009/1001/p692.html

D'Arrigo, T. (2017, September 26). *What Is Polydipsia?* WebMD; WebMD. https://www.webmd.com/diabetes/polydipsia-thirsty

Gandevia, S. C., Allen, G. M., & McKenzie, D. K. (1995). Central Fatigue. *Advances in Experimental Medicine and Biology*, 281–294. https://doi.org/10.1007/978-1-4899-1016-5_22

Holland, K. (2013, June 10). *All About Electrolyte Disorders*. Healthline; Healthline Media. https://www.healthline.com/health/electrolyte-disorders

Jin, W. (2019). Alcohol and Hangovers - Being a Responsible Hedonist. *The New York Journal of Style and Medicine*. https://www.nyjsm.com/Medicine/Emergency/Responsible_Hedonist_JinW.cfm

Marcus, B. H., Williams, D. M., Dubbert, P. M., Sallis, J. F., King, A. C., Yancey, A. K., Franklin, B. A., Buchner, D., Daniels, S. R., & Claytor, R. P. (2006). Physical Activity Intervention Studies. *Circulation, 114*(24), 2739–2752. https://doi.org/10.1161/circulationaha.106.179683

Peacock, O. J., Thompson, D., & Stokes, K. A. (2012). Voluntary drinking behaviour, fluid balance and psychological affect when ingesting water or a carbohydrate-electrolyte solution during exercise. *Appetite, 58*(1), 56–63. https://doi.org/10.1016/j.appet.2011.08.023

Pruna, G. J., Hoffman, J. R., McCormack, W. P., Jajtner, A. R., Townsend, J. R., Bohner, J. D., La Monica, M. B., Wells, A. J., Stout, J. R., Fragala, M. S., & Fukuda, D. H. (2014). Effect of acute L-Alanyl-L-Glutamine and electrolyte ingestion on cognitive function and reaction time following endurance exercise. *European Journal of Sport Science, 16*(1), 72–79. https://doi.org/10.1080/17461391.2014.969325

Quitkin, F. M., Garakani, A., & Kelly, K. E. (2003). Electrolyte-Balanced Sports Drink for Polydipsia-Hyponatremia in Schizophrenia. *American Journal of Psychiatry, 160*(2), 385-a-386. https://doi.org/10.1176/appi.ajp.160.2.385-a

Wiese, J. G., Shlipak, M. G., & Browner, W. S. (2000). The Alcohol Hangover. *Annals of Internal Medicine, 132*(11), 897. https://doi.org/10.7326/0003-4819-132-11-200006060-00008

Wong, S., Sun, F.-H. ., Huang, W., & Chen, Y.-J. . (2014). Effects of Beverages with Variable Nutrients on Rehydration and Cognitive Function. *International Journal of Sports Medicine, 35*(14), 1208–1215. https://doi.org/10.1055/s-0034-1370968

World Health Organization. (2013). Guideline: updates on the management of severe acute malnutrition in infants and children. In *World Health Organization.* https://www.who.int/publications/i/item/9789241506328

World Health Organization. (2018, February 2). *Cholera Infographics.* World Health Organization. https://doi.org//entity/mediacentre/infographic/cholera/en/index.html

World Health Organization. (2021, February 5). *Cholera.* WHO; World Health Organization: WHO. https://www.who.int/news-room/fact-sheets/detail/cholera

www.ingramcontent.com/pod-product-compliance
Lightning Source LLC
Chambersburg PA
CBHW021825190326
41518CB00007B/743